Optimizing High-Performance Teams

A Project Manager Guide

Stephen T Boschulte

Pinnacle Learning Group

Hardback ISBN: 979-8-9999354-9-6
Paperback ISBN: 979-8-9999354-8-9
eBook ISBN: 979-8-9999354-2-7
Audio ISBN: 979-8-9999354-3-4
LLCN: 2025918416

First Edition 9-8-7-6-5-4-3-2-1

DEDICATION

With the advances in technology and service-based industries, the number of programs and projects have increased significantly over the past three decades. Some experts place this year over year growth at over 3% annually. Because there are many teams that make up large programs and projects, individuals can get lost trying to find their path as they navigate their professional journey. This book was written to help provide useful information to those starting on their journey. The author provides tips and techniques learned during over 30 years of professional experience in this space. The lessons learned in optimizing high performance teams in technical program and project management techniques, and the knowledge references provided are useful for anyone along any stage of their journey.

This book is dedicated to those lifelong learners pursuing a better understanding of what it takes to drive high performing teams in consulting and managing technical projects.

ACKNOWLEDGEMENTS

I would like to acknowledge my high-performing support team.

My oldest daughter Lani, whose candid review challenged me to better formulate and communicate my thoughts. The audio book could not have happened without her technical expertise.

My youngest daughter Leila, whose perspective was important to polish the document for the target audience.

My wife, Kathryn, who is my biggest supporter and allows me to pursue my professional endeavors while keeping me grounded.

My support team extends to my mother and brother, the Boschulte, Hariston, and Thomas families and all those who have helped in my professional development over the past three decades. This includes the communities of the Virgin Islands, Columbia Maryland, University of Virginia, University of Notre Dame, colleagues I have worked with at Ernst and Young, Cap Gemini, Deloitte, CACI, and the many clients I have had the privilege of partnering with to meet their missions. Finally, my editorial, advisory and publishing team of advisors who allowed me to make a product that can be communicated to the masses.

CONTENTS

FOREWORD

When someone asks me what are common traits of high-performing technical team members that I manage, I answer with the following 7 traits:

- Strong Work Ethic – someone who takes ownership and is committed to completing their work aligned with the mission.
- Trust: someone who trust that others on the team will stick to their commitments and, likewise, they are hyper focused on sticking to their commitments. They act ethically in all interactions and expect the same for their team mates.
- Enlightened: someone who seeks understanding of the full picture and how their work fits in the overall solution.
- Active Listener: someone who listens at least twice as much as they talk.
- Proactive Participation: someone who actively participates in any meeting and is not afraid to respectfully disagree or share opinions.
- Empathy: someone who seeks to understand others point of views.
- Communication: someone who confirms that the message received by their audience is the message they intended.

Optimizing High-Performing Teams takes this a step further by looking at a standards-based approach to enhance technical programs and projects teams from pre-initiation through post shut down. Tools and tips are provided to enhance project teams across ten project areas

- Program Management Office
- Business Development
- Recruitment
- Integration Management
- Communication Management
- Requirement Management
- Solution Development
- Security Management

- Information Dissemination
- Operations and Maintenance

Project managers of all levels of experience can utilize tips to optimize their high performing teams better through a focus on standards, life-long learning, and a continuous improvement mindset.

Optimizing High-Performance Teams provides a standards-based approach to successfully managing programs and projects. Organized around the project life cycle, this book takes readers on a journey of optimizing high performance teams from program inception to program closure. A clear roadmap is provided, allowing readers to focus on key topics, practical tips, and standard templates that best meet their needs. This guide helps readers:

- Conduct an assessment to develop a personalized roadmap
- Set up and maintain an effective program management office
- Implement leading practices for business development
- Recruit effectively for rapid resource onboarding
- Optimize high-performance teams
- Understand tactical approaches to data migration and system development
- Communicate more effectively with customers and stakeholders
- Manage scope creep
- Execute technical projects successfully
- Consolidate essential tips and techniques for managing projects and programs
- Identify most useful project tools and templates

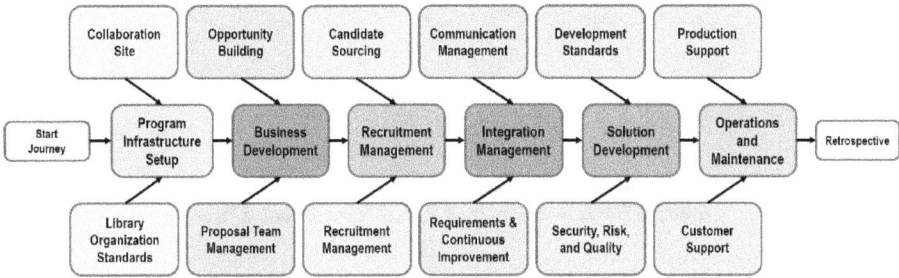

Figure 1.0: Technical Program Roadmap

This book is organized based on the Technical Program Roadmap in Figure 1.0 above and has several types of callout boxes throughout.

Career Note: Considerations for individuals interested in pursuing a particular area of study

Anecdote: Personal scenarios based on practitioners' working experience.

Note: General tip or note about the topic.

1 PROGRAM INFRASTRUCTURE SETUP

The objective of this section is to begin with the end in mind, creating an environment that supports knowledge sharing, continuous learning, clear standards, and procedures that will be scalable to grow and expand with the program.

Primary Goals:

- Where do we begin?
- Program Office Structure
- Team Heuristics
- Library Organization
- Standards
- Measuring Program Performance
- Thank you, Thank you, Thank you

Setting up the infrastructure for effective management of a project management office (PMO) is often more difficult than it seems. On larger programs where this is necessary, the program often has multiple competing priorities to include:

- recruiting the individuals required to perform the work
- creating a collaborative space the team can use to work effectively
- developing a library of standard development lifecycle artifacts and associated processes
- and determining the correct mix of performance and other human resource management to minimize administrative overhead

If the management team is new, an additional opportunity exists for setting up an effective management structure, normalizing and setting expectations, and communicating what each person brings to the team.

1.0 Where do we begin?

Before initiating any new initiative within an organization, it is best to start with an assessment. Individual perspectives contribute to a deeper understanding of how to grow an organization and achieve greater success. Surveys can help identify key areas that require immediate attention, while a strategic plan can ensure the organization, program, or initiative quickly meets its goals and objectives.

Larger organizations often invest heavily in consulting firms to conduct assessments and guide change implementation. One key advantage of using a third party is that employees tend to be more open to sharing candid feedback anonymously, without fear of offending colleagues they interact with daily. However, organizations can also conduct internal assessments effectively—without external consultants—if they ensure autonomy and follow best practices for survey design and data collection.

Below are leading practices for creating and utilizing organizational maturity assessment surveys:

- **Balance focus areas** by limiting categories to 5-7 key topics.
- **Use a variety of question types**, including multiple-select, free-text, and ranking questions.
- **Incorporate open-ended questions** to capture additional details on the survey topic.
- **Limit survey length** to ensure completion within 10-15 minutes and reduce participant fatigue.
- **Start with key business processes** as the foundation for assessment.
- **Engage a diverse group of stakeholders** to gather insights from different perspectives.

For the purpose of this audience, the survey may be conducted to gain a better understanding of each chapter in this book.

The **Information Technology Infrastructure Library (ITIL)** is a widely recognized framework of best practices for managing and improving IT support and service delivery. According to ITIL (2024):

"The implementation and management of quality IT services that meet the needs of the business. IT service management is performed by IT service providers through an appropriate mix of people, process, and information technology."

Although multiple frameworks and standards exist for IT service management, ITIL remains the most widely adopted globally.

Before creating or implementing an organizational assessment, it is essential to understand maturity assessment levels. Selecting the most relevant primary and secondary practices for your organization helps determine its current level of maturity (ranging from Level 1 to Level 5) in the targeted areas. This insight provides a foundation for developing a roadmap focused on key improvement areas. The success of the initiative can then be measured through future assessments.

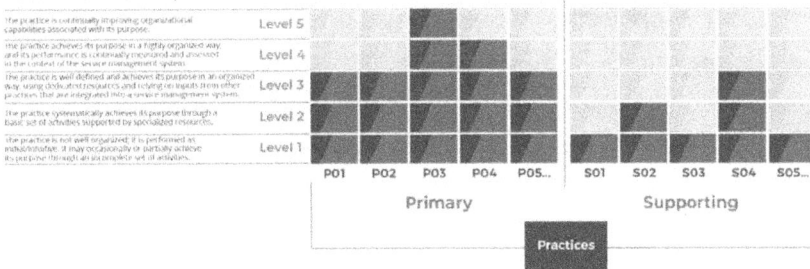

Figure 2.0: Maturity Assessment Levels and Descriptions (Axelos.com, 2021)

Capturing and measuring the maturity profile for each competency area or chapter helps uncover underlying improvement opportunities and identify areas that best align with your organization's goals. **Bi-directional communication** in surveys is most effective when open-ended questions are included for each category (e.g., *What is the company doing well? What can be improved?*).

Effective surveys achieve two key objectives:

1. **Management Commitment** – Proactively capturing and implementing action items based on survey insights demonstrates an organization's commitment to listening to its team and its willingness to accept constructive feedback.
2. **Implementation Roadmap** – Well-designed surveys highlight both short- and long-term opportunities, enabling leadership to make informed decisions on initiatives that provide the greatest value to the organization.

Anecdote: In my career, I have always enjoyed meeting with stakeholders to perform organization assessments. I enjoy hearing from various stakeholders who are knowledgeable about what they do and their commitment to the mission of the organization. Although not all organizations may be in a position to hire a team to conduct such formal assessments, open-ended questions can level the playing field by providing participants an opportunity to provide feedback and anonymous suggestions, if necessary.

If your organization is not already conducting surveys on a regular basis, there is no better time to start than now. If you are launching a new organization, it is essential to incorporate a feedback loop through surveys and assessments as early as possible. Surveys can also be used in conjunction with other management techniques such as **Strengths,**

Weaknesses, Opportunities, and Threats (SWOT) analysis, to gain deeper insights.

For more details on SWOT analysis, refer **to Section 3.1.2**, and for a list of suggested templates to begin your assessment journey, see **Appendix 2: PMO Maturity Assessment.**

1.1 Team Construct

Optimizing a high-performance team requires good upfront planning with an experienced leader who has managed projects of various sizes, types, and complexities. Having someone who is credentialed is preferred, as the most critical skill of a project manager is communication. The biggest benefit of having a credentialed person is that the organization has a high level of confidence the individual has the minimum knowledge necessary to communicate consistently with others with similar competencies. This is true for educational credentials such as a bachelor's, master's, or doctorate degree, as well as professional credentials such as the Project Management Professional (PMP), Certified Information Systems Security Professional (CISSP), Certified Scrum Master (CSM), Project Management Institute Agile Certified Practitioner (PMI-ACP), or even industry credentials such as the Certified Public Accountant (CPA) or the Certified Utility Safety Professional (CUSP), etc.

There are many ways an individual can be credentialed, and every project is different, so understanding the right mix of credentials is the start of optimizing a good team. This is not to say that certified individuals are the only ones who should make up a team, as there are many professionals with life experience and a commitment to lifelong learning who are also effective in growing high-performance teams. Understanding the ideal makeup of the team up front allows you to better assess new candidates. Ensuring the focus is on building a diverse team with unique perspectives will allow the team to be successful through strong and capable project management.

1.1.1 Program Management Office

Regardless of the size of the organization, there are tasks that need to be accomplished to create a program office optimized for efficiency. How does one build an efficient program office, and why does it matter? Projects fail when the management team and program office do not have the necessary support from each other. To ensure success, various activities must be delegated to individuals on the management team, with a minimum of four key roles required to optimize an effective program office.

1.1.1.1 Program Manager

The program manager is responsible for the program. This person is tasked with optimizing a team that achieves success, using time management as an essential skill for overseeing these four primary components:

1. Customer Communication
2. Scope Management
3. Team Execution
4. Contractual Deliverable Submission

All members of the office should report directly to the program manager, with the only exception being when the number of direct reports exceeds ten individuals. Some program offices assign the program manager and a deputy program manager as the Human Resource (HR) manager during the initial hiring period to facilitate administrative tasks related to onboarding new resources. This approach is recommended for larger programs if there is a clear process for transitioning to a dedicated HR manager or onsite manager within two time-reporting cycles.

The program manager should have a clear understanding of all processes in case they need to step in for someone on the management team. For technical programs, it is most effective if the program manager has knowledge of technical solution design, development techniques, and industry trends to enhance customer engagement. Additionally, the program manager should establish a clear continuity plan for when they are out of the office, delegating tasks across various members of the management team.

Implementing this model sets the tone for two important guiding principles in optimizing high-performance teams. First, every team member is responsible for identifying their backup and ensuring the backup is trained and capable of handling tasks during an absence. The designated backup must also take ownership of understanding and executing these tasks when needed. This approach places responsibility and accountability on both individuals, ensuring continuity while reducing the burden on the manager, who should only need to step in for emergency cases.

The second guiding principle is to have every team member create an out-of-office plan. This should be done early in the program, especially for resources who wear multiple hats and cannot transition all activities to a single person. In such cases, an out-of-office plan identifying multiple backup team members is necessary. Each person is responsible for understanding their primary tasks and training their backups on both primary and secondary responsibilities. This minimizes program risk in emergency situations or when individuals leave the program. Additionally, an out-of-office plan supports transition planning. See **Section 2.1.3 *Planning for Leave*** for more information.

Note: During the onboarding session, it is good to provide the rationale of the backup training strategy. One sign that there is an issue in this area is how often team members are called while on vacation. If this happens, it means the program is stuck waiting for a decision and the team member is not given the time to decompress/recharge, which is necessary on high-performance teams.

Anecdote: On one program, a team member went on a well-deserved family vacation during a critical production release. An issue was discovered in the developer's code that required a deep understanding of the code set. When the team called the developer's emergency contact, a family member answered and said, "He will have to call you back in 30 minutes, he is on top of Space Mountain". As the team waited for the return call, stress levels increased because the customer was aware of the issue and tracked the remediation time as part of their quality metrics. In this particular scenario, the problem was resolved within a couple hours but it caused unnecessary stress on the team member, the customer, and the larger program team and interruption to the family's quality time.

> **Career Note**: The program manager should be well versed in program and project management. The person in this role should either already have or set a goal of obtaining, at minimum, a Project Management Professional (PMP) certification, Program Management Professional (PgMP) certification, or an advanced degree in project management. Having a deep knowledge of the business, facilitation and communication, is helpful with the executive management interaction.

1.1.1.2 Deputy Program Manager

The Deputy Program Manager (DPM) position is one of the toughest roles on the program. This role is responsible for much of the day-to-day management of the team and must stay aligned with the program manager's decision-making to step into that role as needed. In cases where there is administrative overhead in onboarding new resources, the DPM facilitates the process by managing the initial onboarding before transitioning new hires to their onsite or HR manager.

Focus areas within the program office can be divided between the Program Manager and Deputy Program Manager based on their backgrounds. For example, financial management, delivery oversight, scope management, or scheduling can be delegated to the Deputy, while the Program Manager focuses on overall scope, contracting, and relationship-building with the customer. The Deputy must be comfortable making key decisions when the Program Manager is unavailable or when they have the designated decision authority for the role. Strong communication, facilitation, and delegation skills are essential, along with a solid competency in the industry.

On larger programs, there may be multiple deputies. When multiple deputies are present, responsibilities can be divided based on the number of resources, specific competencies, or both. If the contract includes both development and infrastructure maintenance, it is best to assign a dedicated deputy for each area. The Deputy focused on development is responsible for solution functional delivery, including capability solutioning and collaborating with the customer on functional requirements. Meanwhile, the Infrastructure Deputy oversees resources managing systems, networks, security, and customer service.

Career Note: The deputy program manager should be knowledgeable in program and project management. The person in this role should either already have or set a goal of obtaining, at minimum, a Project Management Professional (PMP) certification. Project Managers with a deep knowledge of the business, facilitation and communication skills are often successful in this role.

1.1.1.3 Program Support Specialists

The program support specialist focus should be in at least one of four areas:

1. Finance/Contracting
2. Recruiting
3. Knowledge Asset Implementation (Structure/Content)
4. Onboarding/Compliance

If these processes are well defined, it releases a large burden from the Program and Deputy Program managers, allowing them to focus their efforts on the team's health and execution.

> **Career Note**: The product support team member supporting this activity can study and prepare for schedule specific certification (i.e., PMI - Scheduler)

1.1.1.3.1 Finance/Contracting

The finance and contracting tasks require someone with a finance background, a strong understanding of program financing and supporting tools, and excellent interpersonal skills to work effectively with subcontractors and the team. These areas focus on supporting program needs through critical administrative tasks such as managing and filing official contracts and payments, reviewing and auditing all team members' time and expenses, and forecasting resources to meet or exceed program goals.

On larger programs where finance and contracting become more complex due to multiple contracts, subcontractors, and other variables, this area may require a full-time equivalent (FTE) position. Many organizations have standard tools for managing the program and project financials. Understanding the full financial life cycle and how financial data informs both the program and company decision-making is key to successfully managing this area.

1.1.1.3.2 Recruiting

Recruiting, when done correctly, can yield significant long-term benefits. It is often where a candidate gets their first impression of the company, customer, and team. It is crucial to have a resource who is charismatic, people-oriented, and capable of effectively 'selling' the program. Selling the program involves multiple roles throughout the recruitment process, from prospecting candidates to interviewing and selecting team members.

Defining key metrics for each phase of recruitment upfront allows the PMO to quickly assess challenges and implement necessary changes to attract and retain top talent. If an organization conducts exit interviews, incorporating feedback from these discussions can help refine recruiting strategies and improve talent acquisition.

A well-structured recruiting process can be measured over time through metrics such as:

- The average number of days a position remains open
- The performance metrics of employees who went through the formal hiring process
- The retention rate of the organization

Retention rates can be compared against industry benchmarks, such as those published in reports like the *News Release (Job Openings and Labor Turnover Statistics, 2024).* Understanding industry trends in retention helps organizations assess their standing. For example, if the turnover rate in the service industry is 8% and a program has a 6% turnover rate, it may indicate strong retention. However, if the program's turnover rate is 10%—higher than both the company average (6%) and the industry benchmark (8%)—it suggests a problem. In such cases, seeking insights from high-performing organizations can help identify effective strategies for improvement.

Once a well-defined recruiting process is in place, it can be delegated to a junior resource. The person managing recruitment must have strong organizational and interpersonal skills, as they will coordinate with prospective employees, recruiters, management, and customer representatives.

An efficient internal recruiting process begins with identifying key meetings to track hiring progress. Organizations typically hold internal meetings to discuss the status of open positions, as each unfilled role directly impacts revenue. When working with any recruitment team (internal, subcontracted, or external), it is critical to establish expectations upfront regarding:

- Role definitions and candidate qualifications
- The hiring process and responsibilities of each participant (e.g., recruiter, interviewer, PMO, candidate)
- Lessons learned from previous recruiting cycles

A recurring meeting between the PMO and the recruitment team should be scheduled before the organization's broader status meeting to ensure the most up-to-date staffing information is shared. Standardized templates should be developed to track the status of open positions. This process should also include:

- A schedule template outlining dedicated interview slots
- Standard job requisitions (JRs) for each role
- Pre-qualification questionnaires
- Feedback form templates for interview assessments

See **Section 1.1.1.3.4 *Onboarding*** for additional information on techniques for successfully integrating new hires.

1.1.1.3.3 Knowledge Asset Implementation

The individual responsible for managing program knowledge assets must possess strong organizational and communication skills. Effective collaboration begins with structuring documentation so that any team member can locate necessary artifacts, even if they are outside their direct scope of responsibility.

This role is sometimes referred to as the *program librarian* and involves frequent communication with the program office, customer, and project team. More mature programs often assign a technical writer to this position to review and finalize formal deliverables, as these are key contractual artifacts requiring consistent maintenance. Other responsibilities may include managing templates, standards, glossaries, and training materials used across the program.

For larger programs, this role may naturally evolve into managing multi-day onboarding training sessions. See **Section 2.1.1 *Onboarding*** for more information.

Investing time in structuring a well-organized knowledge management system is a valuable efficiency booster. Establishing a standardized documentation framework early on can save significant time and effort later. PMOs can leverage best practices from organizations such as ITIL, PMI, or other standards-based entities to develop a knowledge management structure tailored to their program's needs.

If a new program is being established, dedicating time to structuring knowledge assets before ramping up hiring efforts can prevent inefficiencies and streamline operations.

1.1.1.3.4 Onboarding

The onboarding and compliance area is crucial for maintaining clear communication regarding status updates both internally and with the customer. Once a well-defined process is established, onboarding can be delegated to a more junior resource who is detail-oriented, organized, and skilled in task management.

Starting with the end in mind, key metrics should be captured throughout the onboarding process, including:

- Prospect contact date
- Recruiting response date
- Interview schedule dates
- Acceptance date
- Security forms submitted
- Mandatory training completed
- Access granted
- Onboarding complete date

Given the heightened background check requirements on government contracts, it is important to identify and leverage opportunities to expedite tasks related to the program team. A checklist should be created to outline the specific responsibilities once they are finalized. Below are examples of tasks to include in the checklist:

- Verify candidate meets residency requirements
- Verify completion of background investigation
- Confirm starting date
- Order equipment (laptop, monitor, peripherals, etc.)
- Confirm work location, assign cube, and request a phone number if needed
- Complete badging form and notify export team if needed for foreign nationals

- Send Day 1 email instructions, orientation & Welcome Packet Email
- Open onboarding tickets, obtain approval, and follow up until necessary access is granted
- Process I-9 forms for new hires
- Assign to distribution lists
- Confirm with management approval if working with a new subcontractor
- Coordinate with Finance (for project charge codes)
- Security for new hires (notify management team of new hires)
- Plan and schedule orientation sessions

Note: Some government organizations have a mature process for approval steps to access their systems. If the approval process is not automated or actively managed, an approval task may sit in someone's queue for a long time. Understanding the end-to-end process allows both the customer and the client organization to work together to determine the best process to track the progress and help minimize any schedule risk due to onboarding delays.

1.1.1.3.5 Compliance

The compliance area is another area where you can start with the end in mind. What are the key metrics that the program needs to capture? Understanding mandatory items of interest is an important first step. Service Level Agreements (SLAs) may be officially set by the contract or customer for providing deliverables, annual training, onboarding expectations, etc. From an internal company perspective, there may also be compliance requirements for training, timekeeping, recruiting, and more. Once the key metrics are defined and a solid process is established for capturing them, this area can be delegated to a more junior resource with strong documentation and analysis skills.

1.1.1.3.6 Other Potential Tasks for Program Office Distribution

The project coordinator/support position may also be utilized to perform additional tasks such as:

- Planning and scheduling deliverable sign-off meetings
- Reviewing and preparing deliverables for submission
- Updating project roster, floor plans, and seating assignments
- Gathering quality metrics for monthly reporting
- Managing metrics for onboarding
- Submitting travel authorizations and miscellaneous expense reports for management
- Managing virtual conference accounts
- Managing badge expiration dates
- Planning and coordinating program-wide team-building activities such as volunteer events, social events, etc.
- Ordering supplies, printer cartridges, coffee, etc.
- Handling ad hoc administrative tasks

PMO success is more easily achieved when you have the right mix of support staff to assist the leadership team. Implementing a collaborative infrastructure that includes a library of standard processes and procedures allows the PMO to create supporting documents that communicate to the broader team. This approach helps minimize administrative overhead and maximize the time the team spends on deliverables.

Delegation of tasks is an important skill for everyone on the team. Managers should be provided with guidance on leading practices in this area. Once again, we can turn to our academic community to learn more. Harvard Business School Online publishes their Business Insights Blog (Landry, 2020), which provides tips for new managers.

1. Know what to Delegate
2. Play to Your Employees' Strengths and Goals
3. Define the Desired Outcome
4. Provide the Right Resources and Level of Authority
5. Establish a Clear Communication Channel
6. Allow for Failure
7. Be Patient
8. Deliver (and ask for) feedback
9. Give Credit Where It's Due

Effective delegation is an important skill to have as part of the management team of any size program. Individuals who are placed into management roles usually take on tasks that more junior team members can accomplish to save time. To grow a high-performance team, it is important the leadership team understand their value in communication and decision making and delegating any other tasks to the most appropriate team member.

> **Note**: One approach is to start by delegating as many tasks as you can to the most junior resource on the team. The right junior staff members are often eager to learn and provide value. This approach may take a bit of time; however, it pays dividends during crunch time when the leader has to focus on strategic initiatives.

1.1.1.4 Team Engagement

Optimizing high-performance teams involves creating a culture of engagement. To better understand this concept, we can lean on educational organizations such as the Stanford Center on Philanthropy and Civil Society, which publishes the *Stanford Social Innovation Review* (SSIR). According to Amy Born and Gali Cooks in their article, 'Does Salary Matter?' (Cooks, 2022), there are five key elements that drive engagement:

1. Feeling that the organization demonstrates care and concern for employees
2. Having confidence in the organization's leadership
3. Believing that employee well-being is a priority within the organization
4. Feeling that there is open and honest two-way communication within the organization
5. Feeling like you belong to the organization

The management team must focus on team engagement to optimize a high-performance team and minimize risk. A strong management team needs to ensure they show empathy toward employees. For instance, allowing a flexible workplace where individuals are not tied to a typical eight-to-five schedule is one example. Maintaining a core set of hours for meetings (e.g., 9 a.m. to 3 p.m.) allows for flexibility while keeping team collaboration intact. Management should support employees in their aspirations by offering opportunities for growth through educational programs or job interests.

The management team must also demonstrate a clear understanding of the customer's mission and the confidence to achieve the solutions designed to meet that mission. They need to prioritize employee well-being, encouraging individuals to take time off if they feel unwell or need to visit a doctor. Open and honest two-way communication is essential. Challenges should be shared with the team as soon as possible, when appropriate. An open-door policy should be maintained, with managers setting aside regular time to meet with each team member. This approach helps individuals feel like they belong and their contributions matter.

When the management team can align an individual's personal goals with their professional goals and the program's objectives, it fosters a high-performing team—one that is highly motivated, talented, and engaged in the program's mission. This alignment helps the team exceed customer expectations and fosters respect for each individual on the team.

1.1.2 Tools and Repositories

Mature project management offices reflect the nature of the projects being run. Tools and templates are standardized to maximize program efficiency by addressing the number one goal of project managers: communication. A collaboration site should be created where everyone on the team has access to the information they need to perform their job and communicate status updates in a way that minimizes overhead. The structure of the collaboration site should reflect the team structure, beginning at the highest levels. The site should also include links to training for these tools, and an overview of the training should be incorporated into the onboarding process. For more information regarding onboarding, see **Section 1.1.1.3.4**.

Note: If the program manager has two deputy managers, one for Application Development and the other for Infrastructure and Operations, then it is recommended that you have at least three sections: Program Office (capturing the knowledge assets including templates, training, deliverables, and contract material); Application Development (capturing the development standards and process with deliverables from each application project); and Infrastructure and Maintenance, (capturing the key performance indicators and SLAs for help desk response and infrastructure reference content).

The collaboration site's structure should be standardized specifically for the program. All projects within the program should be managed consistently, with any exceptions to the standard approved and reviewed for continuous improvement. Using this approach results in efficiencies that allow the program office/project management team to easily find backup and surge resources across projects, with minimal disruption.

1.1.2.1 Project Summary Agreement

Scope management and communication of scope are critical processes to standardize at the beginning of any large program. The role of the program office is to support the team, and how the office manages scope with the customer can make or break a project.

An agenda item should be reserved during recurring touchpoints with the customer to review proposed initiatives throughout the project's lifecycle. This open communication helps align the program manager and the customer on potential new work and its impact on existing projects. It also provides an opportunity to capture additional insights into the business value, the likelihood of tasking, and any follow-up actions required by both parties for more rapid approval and execution.

A project approach agreement can be created to facilitate this process. If a formal request for approval on a new project's process is implemented using the organization's template for a new project request via email, the email can be standardized as part of the process for the program management team, or the individuals responsible for requesting and securing approval and funding. The sections of this template may include instructions, complexity, deliverables by project lifecycle stage, and the approval template with signature blocks.

Success in this area is often achieved through standard tools and templates that help the team quickly understand the scope of work required to execute future projects. For mature programs, the program office typically has a template for quickly assessing high-level estimates using a standard set of assumptions and inputs. The goal is to avoid delivering more time than necessary on future projects.

The estimating template should be structured to provide the necessary details for both the program and customer executive management teams. Thresholds can be established for automatically determining which deliverables will be required, the complexity levels, and the number of resources needed to provide a conservative high-level estimate for the initiative. The template should also capture assumptions, risks, and constraints associated with the effort.

Note: Many individuals new to the field of project management find estimating to be frustrating because of the upfront lack of information. One technique is to agree and reconfirm the assumptions of the estimate. The estimates, like project plans and schedules, do change over time. The project team should have a high level of confidence of any estimate or schedule provided to the customer if the assumptions, constraints and risks are constant. This approach makes it more acceptable for the managers providing the conservative input, and allows the decision makers to assess the information to make the best decisions while working with the customer on finalizing the initiative scope.

Many of the assumptions, risks, and constraints are similar across both inflight and proposed projects and can often be reused, with updates specific to the unique items of the proposed project. Like all other processes, tools, and templates discussed in this document, the project estimating tool matures over time, and the team develops a rhythm for quickly assessing the inputs, along with the unique assumptions, risks, and constraints, to provide a conservative estimate that they can confidently accomplish.

1.1.2.2 Project Schedule

Now that the program has a reliable way to manage scope, the next step is to clearly understand the approach of the development team and establish a standard development process for all project teams across the program. This can most effectively be achieved through a standard project schedule, beginning with the project life cycle and work breakdown structure.

It is possible that more than one type of schedule template may be required. For example, you may need different schedules for various types of application development projects, such as transactional system projects, reporting projects, business intelligence development projects, and standard templates for implementing and upgrading new products. Creating standard schedule templates for project managers is the most effective way to ensure consistency in the information provided across the program. The ability to set up this tool for project managers will play a significant role in how easily the program office can expand and contract.

Before the development of any project schedule, there are specific items that should be checked. Below are some of the most helpful tips when using a scheduling tool like Microsoft Project:

- All tasks should be set to auto-schedule to make the best use of the scheduling tool features
- Check to ensure all holidays or nonworking time has been entered
- Confirm the standard completion percentages for consistency across projects
- One approach is to use a limited set of status completion percentages for tasks with definitions such as the following six status levels; 10% - Started; 25%; 50%; 75%; 90%; 100% Completed.
- Determine the default standard days per task type (i.e., 1 day, 5 days, 14 days)
- Finalize the naming convention for all tasks (i.e., begin with verb, first case upper, no duplicates)

- Determine the default standards for internal and external milestones, dependencies, and deliverables (i.e., External Milestones will be in blue text with 'External' at the beginning of task)
- Determine the type and level of resources if resource loading your schedule (i.e., Team or individual)
- Determine the best Work Breakdown Structure (WBS)
- Confirm that all deliverables are called out in the schedule
- Save the name of the file as consistent with the WBS 1.0 Line for each project if creating an Integrated Master Schedule (IMS)

If the program office has a validation checklist that is already in use, the project managers should review the checklist so they can understand, up front, what types of validations are in place as they get started with their new schedule development. See Section 1.1.2.4 Project Schedule-Validation for more information.

1.1.2.3 Project Schedule - Refining for Rhythm

Building the rhythm of a team is one of the most important contributions a project manager can make. A technical development team typically enjoys coding and creating solutions but may be less interested in process and procedures. Understanding how work is being done and communicating how it can be done more efficiently fosters trust and appreciation from the core technical team.

A spreadsheet-based tool can complement the project schedule by setting a repeatable cadence of activity and timing between detailed tasks, which ultimately improves the accuracy of the project schedules. This bottom-up approach is especially useful in setting development standards and heuristics that should be employed by each project manager. By completing this activity upfront, all project managers can develop schedules in a consistent manner that aligns with the integrated master schedule.

In short, the benefit of this process is developing standards and associated rules that the project manager should maintain for high-quality schedules. This in turn reduces the overhead needed to assemble integrated master schedules. Informally setting specific release dates helps with communication across teams. For instance, if all team members know that a formal release occurs on the first day of the quarter, month, or week, and everyone is familiar with their role in the release, the team can begin to self-manage, leading to success. The goal of the project manager is to support the team and ensure they clearly understand and agree to the most efficient development rhythm that leads to proven success.

Having teams who focus on continuous improvement after each release will develop a finely tuned capability at a pace that is ideal for the project team. As the application development team focuses on core development activities, this approach will provide the information necessary to build roadmaps and detailed project schedules that align with the Integrated Master Schedule (IMS). This approach enables a bottom-up/top-down strategy that is continuously updated, determining the approach that works best for the team.

1.1.2.4 Project Schedule Validation

For larger programs with multiple project managers, even if schedule templates are created, it is crucial to confirm the quality of the project schedules generated. In such cases, this role is a good fit for a junior project support team member who is interested in project management. As templates are developed, a checklist of review items should be established, starting with some key characteristics of well-formed schedules. These characteristics may include ensuring there are no ? marks on durations, that all tasks begin with verbs, and that all tasks are resource-loaded.

The validation checklist may vary by program. As the project support team reviews the schedules for consolidation into an integrated master schedule, patterns will emerge, leading to updates to the validation checklist. This checklist should be reviewed with the project management team on a quarterly basis or whenever there are significant changes, based on the observations of the projects submitted by the project managers. The heuristics or rules for developing projects should be included as an agenda item during the onboarding process.

This approach establishes a rhythm for the development teams and projects and enables the program office to drive high value and success for both the company and the customer. For more information on the onboarding process, see *Section 1.1.1.3.4* **Onboarding**.

Having a strong project support team is crucial to the success of a program office. On larger programs, the program leadership team, including the program manager and deputy program managers, must define and implement repeatable processes for the program office. The program support staff must be positioned to learn, implement, and continue executing these processes, enabling the program management team to focus on relationship management, scope management, risk and schedule management, and new business development activities. The tools and techniques provided in this section offer the infrastructure necessary to become a high-performance program.

1.1.2.5 Recurring Meeting Agenda

Next, Effective communication is essential for the success of any program, especially in recurring meetings with customers. To ensure these meetings are productive and contribute to the program's success, the following techniques for meeting facilitation should be employed:

1. **Create an Agenda:**
   Every meeting should have a clear goal, as well as a well-understood purpose and process. This ensures that the meeting remains focused and participants are aligned with the objectives.

2. **Respect Team Members' Time:**
   Begin meetings on time and aim to finish five minutes early so participants have adequate time to prepare for their next meeting. This respect for time enhances the overall efficiency of the team.

3. **Reiterate Next Steps and Responsibilities:**
   If there are action items or next steps, make sure to clearly communicate them and assign responsibility before closing the meeting. This will help ensure accountability and clarity to move forward.

4. **Ensure Active Participation:**
   Encourage participation from everyone by reserving time at the end of the meeting to ask each participant if they have any additional input or questions on the topic discussed. This time

can also be used to confirm the next steps and the expectations for completion.

Establishing clear communication and training for effective facilitation techniques is crucial to optimizing high-performing teams. Communication is at least eighty percent of a project manager's role, and the effectiveness of meetings plays a pivotal part in that communication. Training your meeting facilitators and utilizing the recurring meeting calendar can often be overlooked, but it's a relatively easy area to improve. When done effectively, it can lead to better discussions with the customer and more efficient status reporting. It also fosters candid conversations regarding progress, opportunities, decisions, issues, and constraints.

1.2 Measuring Performance of the PMO

To measure the success of the Project Management Office (PMO), key performance measures should be incorporated into the Service Level Agreements (SLAs). Goals can be established for each metric, with the understanding that these goals will be reassessed periodically. Typically, this reassessment occurs at the end or renewal of a contract.

In mature organizations, however, these goals and metrics are reviewed more frequently, allowing for the identification of any areas requiring additional focus to meet the established objectives. Regular intervals of assessment enable a more dynamic approach to program performance, ensuring that any challenges or opportunities are addressed promptly.

A systematic approach to analyzing metrics maximizes the impact of continuous improvement efforts by focusing on areas that offer the largest value to the company. This approach ensures the PMO remains aligned with program goals and that performance consistently improves over time.

Metric	Description	Goal Setting Criteria
Release Count	The number of software releases over the reporting period.	Increase by a given percentage.
Release Quality	The average of the successful release over the total number of releases.	A minimum quality expectation should be set.
Delivery Count	The total number of solutions delivered on time during the reporting period.	The goal may be a set number or as a percentage of on-time delivery.
Operation and Maintenance Cost	The cost to operate and manage the system over the reporting period.	Reduce the operating cost by a reasonable percentage

Metric	Description	Goal Setting Criteria
Development Cost	The cost to develop the program solutions.	Reduce the average cost of development (i.e., cost per story point)
Technical Debt	The amount of non-functional solution development completed in the reporting period.	Reduce the percentage of non-functional items every reporting period.
Defect Count	The amount of defect remediation remaining, to be completed by the team.	Reduce the percentage of defects every reporting period.
Initial Acceptance	The number of features accepted by the product owner the first time presented.	Set a minimum first-time acceptance threshold.
Service Desk Escalation	The amount of customer service tickets created and elevated to the development team for resolution.	Reduce the percentage of tickets escalated over the reporting period.
Customer Response Time	The amount of time it takes to resolve customer issues	Reduce the average time to resolution.
System Availability	The number of times or length of time a fielded solution was unexpectantly not available to the customer.	Reduce the number of unexpected outages and set minimum availability expectations.
Cost Index	The amount of cost expected over the actual cost demonstrates how well a program estimates the cost of the solution.	Determine the minimum and maximum acceptable thresholds.
Schedule Index	The program's accuracy in meeting the solution delivery schedule	Determine the minimum and maximum acceptable thresholds.
Forecast Index	Demonstrates how well the program forecasts solution development.	Determine the minimum and maximum acceptable thresholds.

Table 1: Service Level Agreements - Program

By selecting the most effective performance measures in advance, the organization can streamline its ticket management system to establish clear rules for capturing essential information. This facilitates the automation of key metrics rollup, which will be used to demonstrate the value the program brings to the organization.

For measurement purposes, contract SLAs can be set in collaboration with the customer, clearly defining the metrics, calculation methods, and expectations. Internal reporting periods may be established on an annual, quarterly, or weekly basis, depending on the contract length and team size. During each reporting period, goals can be set to not only meet, but exceed the expectations outlined in the contract.

Setting key metrics early in the process helps enhance communication and alignment among project teams. By incorporating these measures into the standard status reporting process, the program ensures consistent tracking of performance and continuous improvement.

2 TRANSITION MANAGEMENT

The objective of the Transition Management section is to better understand how to maximize the transition on to and off a program. Effective transitioning allows for a lower knowledge learning curve and minimizes retention risks on the program.

Primary Goals:

- Company Onboard Training
- Program Onboard Training
- Individual Learning Plans
- Transition Meetings throughout the project
- Continuous Self Improvement
- Program Offboarding facilitation responsibility
- Individual Offboarding responsibility

2.1 Transition Management

Transition Management includes determining effective ways to proactively exchange knowledge quickly. There are three situations a program management organization can proactively control associated risks. The first is as resources are being onboarded, the second is to actively determine secondary and tertiary coverage during a leave of absence. Finally, to minimize the risk of losing knowledge, the program management team should have a predetermined strategy for individuals who decide to leave the program.

2.1.1 Onboarding

The primary goal of an effective program management office (PMO) is to provide the necessary support for optimizing a high-performance team. One of the most critical processes in achieving this is onboarding new resources.

Programs operating within the government sector often require extensive background checks and security clearances, which can take anywhere from six weeks to over a year to complete. When designing or refining an onboarding process, it is essential to account for these potential delays. Strong technical or functional subject matter experts—who are accustomed to contributing immediately—may become frustrated if they feel unproductive during the waiting period, leading them to seek opportunities elsewhere.

The solution? A well-structured onboarding process. An effective onboarding strategy should achieve several key objectives:

1. Provides an overview of the program and key stakeholders,

2. Sets program expectations,
3. Identifies items that an individual can do immediately to decrease the learning curve of the program.

An added benefit of getting this right is minimizing the risk of non-compliance and the impact of time the management team and other team members are spending utilizing new resources.

Career Note: Regardless of size, the program manager is usually the most senior resource who handles the day-to-day management of the program. As the team size grows, it is easy to delegate onboarding to someone else in the program office. There is always a balance of how much content to share with a new resource and how soon to provide the content to allow the new resource to gain the necessary knowledge for their role. The program manager should make it a point to meet with every resource who is part of the team. On larger teams, one time management technique is to schedule monthly thirty-minute onboarding sessions. The goal of the thirty-minute session is to start building an open-door policy where individuals are comfortable with the feedback. As a program manager, the initial orientation meeting is your chance to make sure everyone hears consistent messaging on what drives success. This is useful if action is required later for resources who are not a good fit to the organization. By having onboarding classes or cohorts, you start to establish an informal relationship where each cohort gets familiar with others across the program. This allows them to get each other up to speed, or to facilitate communication long after the resource is onboarded, thus helping to optimize a high functioning team.

2.1.1.1 Introduction

Starting the onboarding session on time is crucial. This first formal meeting between the program manager and the team sets expectations immediately. Respecting each other's time is a simple yet powerful rule that contributes to optimizing strong, high-performing teams.

During this meeting, the program manager should facilitate personal introductions across the team. This can be as simple as providing a brief professional background, outlining goals, and setting instructions. New team members should also be encouraged to share information about themselves, such as their previous positions, areas of expertise, and what they are looking forward to in joining the team.

If other members of the management team are attending, this session provides a great opportunity for them to introduce themselves in an informal setting. If a deputy program manager or another management team member has a direct report joining a larger program, they may participate for the first five to ten minutes for introductions before moving on to their other responsibilities.

> **Note**: Everyone on the management team is aware of the content as all resources go through this session. This session complements the onboarding done by each of the HR Managers and team leads.

2.1.1.2 Program Overview and Key Stakeholders

At a minimum, the program manager should provide a brief background on the customer and the internal company structure. Links or documents with additional details should also be shared to help new team members explore the information further. This overview helps ground team members with diverse backgrounds and ensures they understand the program's context.

While some of this content may be covered in general company or customer onboarding sessions, hearing it from the program manager's perspective allows new team members to better grasp the program's mission and its role within the company, the customer organization, and the broader industry.

The primary program goal and mission should be clearly stated, ensuring alignment on key objectives and milestone dates. Goals should be specific and measurable; examples may include migrating to the cloud within eight months or delivering a working solution by year-end.

A list of key program stakeholders should also be provided to new team members.

Anecdote: One technique for on-the-job training is to have the new individuals sit in on meetings. Providing the list of important stakeholders up front allows the new team members to understand the roles of the influential and authoritative individuals on the program. The important stakeholders should include persons from the customer and internal program team.

2.1.1.2.1 Customer Stakeholders

The customer stakeholders should include the project sponsor and their up-chain management team. On government programs, this may include the Contracting Officer Representative (COR) and Government Technical Lead (GTL). When applicable, the project sponsor's direct reports and other influential people on the customer and customer's clients should also be included. This may include sponsors of related programs, customer assigned team leaders, and customer assigned process leads such as the leads for the Agile or system development life cycle (SDLC), change management (CM), quality assurance (QA), risk management (RSKM), and release management (RLSM) processes.

2.1.1.2.2 Company Stakeholders

A list of company stakeholders should be provided with emphasis on the management team. This is the opportunity to provide the level of authority management has over the team. If there is a question and the person's primary manager is not available, then the team will have a list of individuals they can reach out to for decision making or communicating issues. From a program perspective, this list will include the program's up-chain management team, such as the Director, Vice Presidents (VPs) and the Executive team as well as the deputy program manager(s), and the program offices support staff. The primary team leaders representing the Business Analyst (BA), System Architect (SA), Data Architect (DA), Development (DEV), Testing (TST), and other team members who may be important for the specific person or class participants can also be included for reference on the list.

2.1.1.3 Program Expectations

With the primary stakeholders identified, the next step in the onboarding process is setting clear expectations. The goal is to ensure that all team members understand the norms, standards, and best practices of the program. Since individuals come from diverse professional backgrounds and management styles, this step helps everyone align with the processes and procedures necessary for success.

Key areas to highlight include:

a. **Communication**: There should be no surprises; team members should raise risks early and often. If a direct manager is unavailable, team members should proactively contact another member of the management team.

b. **Awareness:** The management team should always know a team member's availability and how to contact them in case of an emergency. If someone becomes available unexpectedly, they should inform their manager, another management team member, or the program manager directly. This reinforces the importance of independent time management and helps ensure continuous progress across all areas.

c. **Core Hours**: Set the expectations provided by the customer or establish some of your own. One example is to include a smaller range to provide additional flexibility for the team. Be clear on what this means. For example, everyone on the team is expected to be online during the core hours of 9am to 3pm so flexibility can be driven by having someone work 8am-5pm, 6am-3pm or 9am-6pm. Reemphasize the importance of being on time for all meetings.

d. **Compliance:** Provide clarification on some of the important compliance areas such as annual training, timekeeping, or other compliance requirements and provide the rationale for each.

Anecdote: One exercise technique is to provide an example of how a particular compliance item affects the program. As an example, at one company, timekeeping compliance was always mentioned at every meeting, even by senior levels of the organization. An approach is to break down the cause and effect of one person missing submitting their timecard on time. The individual may say "oh well, I forgot". That one decision led to a chain of events including an automated set of one to four system email up-chain. These emails then triggered an email reaction back down chain leading to up to four additional emails. When you magnify that by a large team, the individual gets to understand the importance of a simple item like timekeeping. Not having the timecard completed at the end of day causes a waste of at least 5-10 minutes per individual, not including the amount of time spent from each executive team member during every meeting.

2.1.1.4 Administrative Items

Onboarding is the best time to take care of general administrative items. Understanding team norms and tips to ease the new resource onto the team is beneficial for both the new team member and the program office. Items in this section may include general process overviews like exception processing, timekeeping, and status reporting, as well as more specific items such as: Setting up email signatures, voicemail, and out-of-office messages, guidelines for handling sensitive information and email attachments, and expectations for cleanliness in the common areas such as breakrooms and workspaces. This section may also include time management tips, such as downloading a timekeeping application or virtual meeting applications to smartphones for quick updates.

2.1.1.5 Self-Study Opportunities

Many organizations have a wealth of knowledge databases available for their employees. Some use well-known web-based providers such as Lynda.com or partner with other providers like Udemy, Edmodo, or Coursera. Identifying a way to enrich your organization's digital library with the appropriate web-based training opens up a world of opportunities for your team.

In situations where new resources are getting up to speed and waiting for a clearance, it is beneficial to encourage them to build the skills necessary for success. Examples include:

- **Project Managers** attending a session on facilitation
- **Developers** reviewing development standards or learning an unfamiliar technology
- **Business Analysts** studying regulatory requirements for their area
- **All Team Members** reviewing a glossary of program acronyms and definitions, assessing role-based certifications, or catching up on industry-leading practices and events

The key message is that there are many ways for individuals to build their professional skills efficiently, making the best use of their time.

Anecdote: Some organizations provide libraries of virtual books and training that are rarely utilized. One technique used to help show the value of this resource is to identify specific skills for each role and to create a slide in the onboarding deck that demonstrates how many books/webinars are available to the new employee regardless of role. For example, if your organization utilized Agile for software development, one should include a metric of how many search items result from querying 'Agile'. Repeating this with four to six targeted competencies can quickly demonstrate the value of these libraries. If your organization has certified individuals, utilizing these inhouse resources to benefit for the certification's continuous learning requirements can significantly reduce the cost of maintaining certifications.

2.1.1.6 Get Involved

Team members in large consulting organizations are often so consumed with project activity and customer focus they feel they have no time for corporate initiatives. However, corporate initiatives should be embedded as part of a performance feedback loop.

Engagement in corporate initiatives provides:

- A better understanding of the organization's internal support structure.
- Opportunities to expand internal networks and connect with professionals from different backgrounds.
- Access to mentors and colleagues who can serve as a sounding board for ideas.
- A platform to showcase leadership skills and contribute to internal projects

Most importantly, corporate initiatives offer team members an opportunity to grow within the organization while strengthening their professional relationships.

> **Anecdote**: Identifying opportunities to share or gain additional knowledge may include reaching out to another team that has a basic training session, facilitating a meeting that demonstrates your area of expertise to others, or sharing lessons learned. This approach enhances one's perception of a knowledge expert in that field and is invaluable for someone just starting their journey.

2.1.1.7 Additional Content

Additional content may be included to emphasize specific items to newcomers so that they are aware of and avoid recurring incidents. Such items may include reminders about interacting with the customers, dress code for office visits, and facilitation tips. This approach makes the onboarding presentation used by the program manager a living document that is updated for each session, and that can act as a regular reminder, if used as part of a regular occurring session with all the program or organization employees.

Reference material often provided for the individuals include:

a. Standards Documentation
b. Glossary of Terms/Definitions
c. Open Positions and Referral Database
d. Productivity tips
  i. Updating picture profiles in email or virtual meeting applications

  ii. Facilitating Sessions

  iii. Process for downloading tools/applications to smartphone

 e. Overall timeline with high-level milestones (if not included in main section)

The onboarding process for any program serves as the first impression for new employees. Formal onboarding sessions can range from a single-day meeting with one or two sessions to a multi-day workshop featuring multiple sessions focused on key program areas. Regardless of the format, the program manager and deputy program manager should attend at least one session with every team member.

Continuous improvement in the onboarding process can be achieved by encouraging new team members to document any challenges they encounter or suggest enhancements as they progress through the process. Introducing this task during orientation fosters a continuous feedback loop, allowing for regular review and refinement of the onboarding process throughout the program lifecycle.

2.1.2 Employee Appreciation

Employee appreciation is one of the most overlooked aspects of optimizing high-performance teams. When implemented effectively, especially when combined with team-building exercises, employee appreciation becomes one of the most impactful yet cost-effective methods for fostering strong, high-functioning teams. Specific steps that can be taken in this area include:

- Identify what opportunities are available for employee appreciation by the organization, division
    - Quarterly bonus
    - Program appreciation

- o Customer appreciation
- Identify the team's interest in opportunities to showcase talent
  - o Allow each individual to provide self-input assessment. This allows the management team to quickly assess perceptions across the team. Questions like who else on the team are role models and why? Provide insights based on feedback from different roles allowing additional insight that is more valuable as the team size grows.
  - o Make sure the management team actively meets with and documents what each direct report has done well, areas for development, and interests.
  - o Customer Support may have better opportunities to get kudos from customers. All feedback should be captured and acknowledged across the team
  - o Technical Support team is a thankless job as consumers are usually only visibly aware when things go wrong. Highlighting successes on major upgrades, new capability, and availability allows you to acknowledge these types of teams
- Certificates of appreciation are an easy way to acknowledge good work
- Team building activities at major milestones. Take some time to celebrate the wins and avoid getting into the habit of what's the next big milestone too quickly. Include a team interaction rule to wait at least 24 hours after a major milestone before inquiring about the next milestone.
- Bragging rights
  - o High-achieving individuals are often naturally competitive. A simple competition between divisions or teams can foster camaraderie, encourage engagement,

and strengthen team dynamics. Additionally, incorporating friendly competition into sports or fitness activities promotes participation in exercise, leading to better health outcomes while reducing stress and absenteeism. Inviting families to these activities provides an opportunity to connect outside of work, strengthening both personal and professional relationships. Recognizing shared interests beyond the workplace can be especially beneficial during challenging times. Furthermore, 'bragging rights' earned through these competitions contribute to friendly office banter, which can serve as a valuable tool for maintaining a positive and balanced work environment, particularly in high-stress settings.

- Providing monetary bonuses is good, however, not always necessary
  - Some organizations offer opportunities for periodic bonuses, yet, I have been on teams where these bonuses were not requested. In other cases, I have seen upper management get bonuses and some were 'more open' about sharing their bonus pool while others keep the pool to themselves. A manager's style usually catches up to them in the long run.
- Making the hard decisions and cutting underperforming resources is an important skill
  - Performance Improvement Plan - gives a chance for individuals to turn their performance around. Make the candidate aware of where they are not meeting expectations and identify a plan to improve. I have had several individuals on these plans that, once made

aware, turned their performance around and became one of the stronger team members.

- In larger organizations, some people continue to try to find ways to 'game the system'. These individuals bring down the high performing team and, if not remedied, leads to less-than-optimal behavior by others on the team. Document and communicate expectations early and often so there are no surprises when decisions are made.
- Understanding the core skill sets, strengths, and weaknesses of an underperforming team member provides a unique opportunity to help that person grow professionally. It may be as easy as making them aware of their shortcomings, having them work with a mentor, or attending formal training. Whatever approach is used to improve the team member's performance should be time specific using SMART goals for a period of no more than three months depending on the deficiency and increased stress on the team based on those deficiencies.

In summary, building trust within a team begins with leadership and relies on each team member's strong work ethic and mutual respect for one another's time and expertise. Team members must trust that their responsibilities will be managed in their absence, allowing them to rest and recharge, ultimately maximizing their contributions upon return. Additionally, they should have confidence that their teammates will fulfill their roles, ensuring collective success. Most importantly, a high-performing team fosters an environment where individuals feel safe to challenge ideas, educate one another, seek understanding, and continuously grow as lifelong learners.

2.1.3 Planning for Leave

Proactively planning for leave is one of the most overlooked opportunities in a program office. Implementing a structured process that can be applied throughout the program lifecycle will offer several key benefits:

1. Team Knowledge Growth: Clearly defining primary and secondary resources for each role ensures continuous development even when the program is in a steady state. The secondary resource can periodically take the lead on tasks, allowing the primary resource to provide coaching and refine existing procedures. This approach fosters ongoing learning and supports continuous improvement.

2. Emergency Preparedness: Establishing a transition plan for leave while the team is operating in a steady rhythm enables the organization to handle unexpected events with minimal disruption. When a leave plan is already in place, team members can rely on a checklist to ensure seamless coverage, allowing individuals to manage personal emergencies with confidence. The designated secondary resource can step in effectively, reducing both risk and stress for the entire team.

3. Reduced Retention Risk: Proactively managing leave also helps mitigate retention risks. High-performing teams typically experience lower turnover rates compared to the broader organization and industry. Team members who are driven by shared goals, continuous learning, and a commitment to exceeding customer expectations are more likely to stay engaged. When a team member departs, whether due to

retirement, promotion, or new career opportunities, it can serve as a testament to the team's effectiveness. Additionally, ensuring the right team fit by addressing underperformance contributes to long-term stability and success.

2.1.4 Offboarding

All programs eventually come to an end, and understanding the specific actions required for an effective transition is essential to ensuring continued success for the customer. Proper planning for the time needed to off-board resources and close the program office is critical. This approach allows the company to maximize customer relationships, maintain a positive stance toward employees, and continuously enhance its knowledge base by capturing leading practices. Conducting a retrospective on what worked well and identifying areas for improvement will help refine strategies for future programs.

2.1.4.1 Individual Offboarding: Team Member

There are several reasons why a resource may be off-boarded. A customer may request removal due to perceived inadequate performance. The sponsor may decide to change direction. A team member may have found another opportunity. Successful delivery of a solution brings the program to an end. Although the amount of effort spent by both the individual team member and the program office will differ depending on the situation, there are key steps that should be followed by the individual.

1. Create a detailed transition plan. See **Section 2.1.4.2 Detailed Transition Plan.**

2. Complete a retrospective of your time on the project/program
    i. What worked well? What didn't? What tips can you provide to make the next person in the role successful?
3. Identify an executive coach or mentor who can help with updating your professional profiles/resumes
4. Update your resume within the first week of notification of off-boarding.
5. Reach out to your network
    i. Internal: If the company has a recruiting/redeployment group, schedule a meeting with your point of contact
    ii. Internal: If there are departments or executives that you either have worked for or would like to work for, now is a good time to check those areas
    iii. External: Referrals are the best way to get the next good position. Your network knows your work, what you are good at, and what opportunities may be available with someone with your expertise
    iv. External: Update your profile on appropriate social platforms and set up alerts for positions that you are interested in. Confirm that you joined the applicable groups where individuals with your professional interest form online communities. This could be industry groups, groups with the same role/competency, or vendor-based groups.
6. Self-Care: Continue your personal retrospective on your work life and personal life and use the opportunity to continue professional growth. Is there a tool that you wished you knew better? A certification you were interested in obtaining? Now is the time to do the research as you continue your job search. It is good to set aside a specific set of hours to job hunt, self-improve, and relax every day.

> **Note**: It is the responsibility of the individual team member to proactively manage their career and there should be no expectation of additional support. Some companies handle transition support better than others.

2.1.4.2 Detailed Transition Plan

A detailed transition plan minimizes the risk of changes on a program if done effectively. A template should be easily available for the team member to complete. The plan should include:

a. Overview/Instructions: Provide clear instructions for the team member on how to complete the plan template and save the file using the standard naming convention. The file name should follow a consistent format and include the name or identifier of the person completing the document. Clear and concise directions ensure the user provides the necessary details required by the program and team.

b. **Transition Plan:** Include a section detailing tasks, the individual to whom each task is being transitioned, the date and time of the transition meeting, and any additional comments.

c. **Recurring Meetings:** Identify recurring meetings attended by the individual to ensure that key meetings continue to be covered. If the resource facilitates meetings, ensure those meetings are reassigned to someone remaining on the program or that the meeting series is canceled as needed.

d. **Documentation:** Specify where company and client working artifacts/documents are stored. These may include formal deliverables, customer working documents, internal company records, and business development materials. Since different types of documents may be stored in various locations, providing direct links will help mitigate the risk of losing access to critical resources.

e. **Open Tickets:** If the customer or company uses a ticketing system, list all open tickets initiated by or assigned to the transitioning resource. Clearly designate a new owner for each ticket. Systems such as help desk, risk management, development, requirements, configuration, and test management should all be included in this list. Ensure that all open tickets are reassigned to the appropriate transition target.

f. **Lessons Learned:** Capturing lessons learned from departing team members helps them reflect on their experience, provides valuable insights to those taking over their responsibilities, and allows the program to act on specific feedback for improvement. Often, meaningful opportunities for growth and refinement emerge only when team members take time to proactively reflect.

g. **Other/Reference:** Additional sections may be included as part of the transition document based on the specific needs of the individual or program. This customization allows team members to take ownership of their document and highlight what they consider most critical to the program team. Any additional sections should be clearly documented in the Overview/Instructions section to ensure the program office understands the purpose.

2.1.4.3 Retrospective

Those who embrace the Agile mindset allow it to extend beyond software development, influencing all aspects of their work and professional growth. Rarely does an individual have the luxury of time to reassess and enhance their professional development, regardless of their career stage. Retrospectives should be conducted periodically and should become a personal requirement during significant life changes. Expanding the concept of a retrospective to include personal life allows individuals to reassess their priorities and long-term goals.

Each team member is responsible for proactively managing their own growth. A strong Program Office can provide critical support during these often-stressful career transitions. By leveraging the same concept and template used for Agile retrospectives, individuals can modify the approach to take stock of their work and contributions on past assignments, identify available tools to focus on their next growth opportunity, and reflect on their career trajectory. Based on this assessment, they can determine specific actions needed to reach the next milestone in their professional journey. Additionally, this reflective process helps broaden perspectives and allows individuals to explore the best options for achieving their aspirations.

2.1.4.4 Mental Health

Personal retrospective analysis begins with cultivating the right mindset. Programs and projects evolve, and some of the most valuable lessons come from setbacks. As part of professional growth, individuals should always strive to do their best and remain confident in their abilities, even when transitioning away from a project.

Certain situations require deeper self-reflection than others—whether a project ends successfully or an individual is reassigned due to performance challenges. How professionals approach their work and engage in their responsibilities significantly influences their ability to navigate inevitable career changes.

2.1.4.4.1 Positive Thinking

Optimism is a common trait among individuals who drive success and make the impossible happen. A positive mindset plays a crucial role in reducing stress and improving overall well-being. According to the Mayo Clinic, individuals can learn to overcome negative self-talk through practice. In their article, *"Positive Thinking: Stop Negative Self-Talk to Reduce Stress"* (Mayo Clinic Staff, 2023), clinical experts discuss the health benefits associated with maintaining a positive outlook.

If a team member tends to see the glass as half empty rather than half full, encourage them to read the article and practice shifting toward an optimistic mindset. Promoting outside reading and self-improvement not only contributes to the development of high-performing teams in high-stress environments but also supports personal growth throughout life's journey.

Note: Enough cannot be said about formal and informal mentors that shape the person you become. Learning via internships and from individuals who are good at what they do is always best. Determine interests that you would like to pursue. Understand who are the good role models and why. Reach out and ask for advice on being successful in that industry. These formal and informal mentors like to share what they are passionate about and will often take the time to give back and teach others coming up in their field.

2.1.4.4.2 Amicable Separation

There are many reasons an individual may separate from a program. The contract may be coming to an end, the individual may wish to explore new opportunities for professional growth, they may not be the right fit for the team, or they may struggle to meet the expectations of the role. Regardless of the reason, it is always beneficial to part on amicable terms. As a team member, the best approach is to consistently give your best effort, learn from each experience, and use those lessons to refine your skills and professional approach. Career growth is a personal journey, and only you can determine the path that aligns best with your goals and values.

2.1.4.4.3 Self Reflection

Professional growth is most successful for individuals who embrace lifelong learning, regularly engage in self-reflection, and consistently ask themselves:

- What could I have done differently?
- How can I improve?
- What areas would make me more effective?
- Are there new skills I should explore based on insights from my last project?

Self-reflection fosters continuous improvement, leading to more focused professional development and, in turn, a more effective and productive team member. Some general areas of knowledge that can contribute to growth include:

- Studying and exploring business or project management competencies that align with personal interests

- Reviewing shortcut keys for commonly used documents, spreadsheets, and presentation tools to enhance efficiency
- Reading industry articles to stay informed on current challenges and emerging trends
- Attending conferences to gain insight into upcoming trends and identify areas for further study

Anecdote: Having a strong support system is critical in high performing team members. Before my marriage, I read so many books to advance my education that the thought of 'reading for pleasure' never crossed my mind. After 'encouragement' from my wife during our limited vacations, I tried it and it opened a whole new world for me. As a result, I now incorporate reading as an important part of my sleep preparation routine.

2.1.4.4.4 Read, Read, Read

Many people underestimate the power of a good book. Those who perform well on high performing teams are also avid readers. According to Healthline (Rebecca Joy Stanborough, 2019), six benefits of reading are:

- Improves brain connectivity
- Increases your vocabulary and comprehension
- Empowers you to empathize with other people
- Aids in sleep readiness
- Reduces stress
- Lowers blood pressure and heart rate

During the transition is a good time to get into healthy reading habits while catching up on the latest trends in one's profession.

Career Note: There are many websites that offer free or low-cost training of general, technical, or project management skills to prepare for success. Examples are provided in the table below.

Category	Title	Company	Est. Hrs	Link
Time Management	How To Use **Google Calendar** as a Planner - MASTER Your Time Management & Scheduling Skills	You Tube	0.5	https://www.yout ube.com/watch?a pp=desktop&v=iW 1zKiKKXjA
Time Management	The ULTIMATE **Google Calendar** Planner System for EVERYTHING | TUTORIAL	You Tube	0.1	https://www.yout ube.com/watch?a pp=desktop&v=dM AFJRml4Oo
Project Management	Microsoft **Excel** - Ultimate Introduction to MS Excel	Udemy	1	Free Microsoft Excel Tutorial - Microsoft Excel - Ultimate Introduction to MS Excel | Udemy
Project Management	**Excel** Beginners Training from Scratch	Udemy	3.5	https://www.ude my.com/course/ex cel-beginners- training-from- scratch/

Category	Title	Company	Est. Hrs	Link
Project Management	Create a **project** in Project Desktop	Microsoft	1	https://support.microsoft.com/en-us/office/create-a-project-in-project-desktop-783c8570-0111-4142-af80-989aabfe29af
Project Management	**M.S Project** for Beginners	Udemy	2.5	https://www.udemy.com/course/ms-project-for-beginners/
Project Management	What Is Agile Methodology? | Introduction to Agile Methodology in Six Minutes	You Tube	0.1	https://www.youtube.com/watch?v=8eVXTyIZ1Hs
Project Management	**Agile** Development with Scrum of Scrums and Scrum Master	Udemy	2	https://www.udemy.com/course/agile-development-with-scrum-of-scrums-and-scrum-master/
Project Management	JIRA CRASH COURSE for Beginners | Jira Tutorial | Jira Training | JIRA Project Management	You Tube	0.5	JIRA CRASH COURSE for Beginners | Jira Tutorial | Jira Training | JIRA Project Management (youtube.com)

Category	Title	Company	Est. Hrs.	Link
Project Management	HOW TO USE JIRA | Free Agile Project Management Software (Jira tutorial for Beginners)	You Tube	0.3	HOW TO USE JIRA | Free Agile Project Management Software (Jira tutorial for Beginners) (youtube.com)
Project Management	Learn **Jira** Complete from Scratch to Expert	Udemy	5.5	https://www.udemy.com/course/learn-jira-complete-from-scratch-to-expert/
Project Management	**Get the most out of Jira** Jira is a widely used software development tool developed by Atlassian, primarily designed for project management and issue tracking.	Atlassian	1.3	Get the most out of Jira : Atlassian

Category	Title	Company	Est. Hrs	Link
Project Management	**Get the most out of Confluence** Confluence is a software that lets you create, share, and harness knowledge across teams with dynamic Pages, Whiteboards, Databases, and videos.	Atlassian	2	https://university.atlassian.com/student/path/861302
Programming	Java for Absolute Beginners	Udemy	9	https://www.udemy.com/course/java-for-absolute-beginners-c/
Programming	R Basics - R Programming Language Introduction	Udemy	4	https://www.udemy.com/course/r-basics/
Programming	Python For Beginners	Udemy	4	https://www.udemy.com/course/python-for-every1/

2.1.4.4.5 Journaling

Journaling is another technique that can be therapeutic and help in long term reflection and help personal development growth. It is hard for some to begin the journaling process. If journaling is being explored, asking the following questions may help initiate the process.

- What is working out well for me?
- What am I most looking forward to?
- What was one thing that made me smile or laugh?
- What was something I learned?
- What was something I accomplished or would like to accomplish?
- What was most challenging?
- How did I make someone's day?
- What is an ideal day from the time I wake up until I go to sleep?
- What is going well and where am I having difficulties? What specific action can I do to keep doing the things that are going well and improve the areas of difficulties.

2.1.4.4.6 Opportunities Abound

Remember that every individual is unique, bringing their own perspective and background to any role. As part of your retrospective process, take the time to understand your personal value and identify ways to enhance it for your next opportunity. During the transition period, set a structured schedule to focus on the following three key areas:

1. Spring Cleaning: In this context, spring cleaning refers to organizing and refining your files from previous projects as you finalize your detailed personal transition plan. This ensures a clean slate for your next endeavor while preserving valuable insights and materials.
2. Preparing: Dedicate time to meeting with a mentor and/or executive coach, updating your resume, and pursuing training in areas where you want to strengthen your skills before your next opportunity. Proactively investing in your development will position you for success in your next role.

3. <u>Self-care</u>: Use this transition period to do something you've always wanted to but never had the time for. Whether it's taking a trip with your family, enjoying the outdoors, painting, photography, reading, or simply catching up on sleep, prioritizing self-care helps recharge your energy and mindset for the next chapter in your career.

Because each person is different, assuming a two-week notice period, one example of a transition plan is shown in Table 2 below.

Week 1		
Activity	Day	Time
Exercise Routine	Daily	7am-9am
Draft Transition Plan/Meetings	Monday/Wednesday	9am-12pm
Schedule meeting with mentor/coach	Tuesday	1pm or 4pm
Update internal and external resumes, sign up for alerts	Tuesday/Thursday	10am-12pm
Review internal jobs	Tuesday/Thursday	1pm – 3pm
Take children to the library	Daily	4pm – 5pm
Family Dinner	Daily	6pm – 7pm
Family Entertainment (Favorite TV shows)	Daily	7pm – 9pm
Bedtime routine (includes reading a book)	Daily	9pm -10pm

Week 2		
Activity	Day	Time
Exercise Routine	Daily	7am-9am
Modify/Finalize Transition Plans/Meetings	Monday/Wednesday/Friday	9am-12pm
Meet with recruiters	Tuesday/Thursday	1pm or 4pm
Explore available options and create decision matrix	Tuesday/Thursday	10am-12pm
Review internal jobs	Tuesday/Thursday	1pm – 3pm
Take children to the library	Daily	4pm – 5pm
Family Dinner	Daily	6pm – 7pm
Family Entertainment (Favorite TV shows)	Daily	7pm – 9pm
Bedtime routine (includes reading a book)	Daily	9pm -10pm

Table 2 - Sample Two Week Transition Schedule

Creating a schedule up front allows you to create SMART goals to achieve the best outcome for you and your next role while providing additional benefits to allow you to recharge and get ready for the next big opportunity. Setting a routine is a key to success in this area.

2.1.5 Individual Offboarding: Program Office

How a Program Office reacts to offboarding is critical to the program. Depending on the reason for offboarding and the size of the program, the program office should do what it can to retain good staff. Change is one of the most stressful times for team members and not having stability increases anxiety levels for everyone involved; including the person affected, the manager, and the project team members who have built solid relationships on high performance teams. The steps that the program office team should be performing include:

1. Meet with the resource and provide the specific reason why they are being offboarded
2. Provide the offboarding template on day one or day two requiring completion within two days
3. Draft an email to send out internally to identify retention opportunities within the company
4. Meet with internal redeployment team
5. Review redeployment plan with resource as time permits … provide recommendations and next steps (see **Section 1.2.4.1 Individual Offboarding: Team Member**)
6. During the transition period, there should be at least 3 meetings
    i. Initial meeting to discuss decision with team members (item #1)
    ii. Meeting to review the initial draft of the transition plan (item #4)
    iii. Final transition meeting to review/confirm that all items are adequately transitioned on the second to last day of the transition period. This approach allows for any last-minute updates prior to departure.

> **Note**: Preparing an out of office plan for resources ensures that the program minimizes risks in changes. A good out of office plan for a team member is one of the best tools to prepare for a transition. In particular, if the program has an expectation that the individual team member is responsible for identifying a backup while they are out of office.

2.1.5.1 Project Coordination - Offboarding

There is a list of specific steps that the coordinator completes as part of an effective offboarding process.

- Open Offboarding tickets, obtain approval, and follow up until access is disabled
- Coordinate asset collection such as equipment, badges, and keys
- Confirm company processes have been completed by managers
- Submit requests to disable email credentials, tokens, etc.
- Remove from distribution lists, collaboration site contacts page
- Confirm any meetings scheduled by the resource are deleted/rescheduled
- Confirm final timecard is completed and signed
- Confirm the individuals have all the contact numbers they need
- Collect assets

2.1.5.2 Initial Meeting with Affected Resource/Team

The first step in managing a significant program change is to determine the messaging that will be communicated to the affected team or individual and to establish the most appropriate communication plan.

Once this is finalized, it is recommended the meeting take place at the end of the workday, preferably on the last workday of the week. This timing allows the individual time to process the difficult message and be in a better state of mind to discuss next steps early in the following workweek.

It is important to recognize that individuals often go through several stages of grief or anxiety before they can productively move forward in seeking their next ideal position. The sooner they work through these phases, the more productive they will be in securing a new, more fulfilling role.

Adequate time should be allocated for the initial discussion. If there has been regular feedback between the individual, the team, and management, the message should not come as a total surprise. The meeting should address the following key points:

1. Purpose: Why are we here?
2. Background: What brought us to this point?
3. Duration: How long is the transition?
4. Next Steps: What are the specific action items following the meeting?

The individual will focus on completing offboarding documents, updating their company profile, and working with the redeployment specialist. Meanwhile, the manager, program team, or company, will begin communications with the team, provide recruiting support, and prepare reference letters. With this structured approach, both the program and the affected team or individual can immediately begin their respective transition activities, making the best use of the available timeline.

2.1.5.3 Offboarding Template Review

A template with clear instructions should be provided to guide the individual through the offboarding process (see section 1.2.4.2 *Detailed Transition Plan*). The template should be reviewed with the individual, offering an opportunity to clarify any required information and emphasizing they can modify the template as needed to ensure all necessary details are captured.

As part of the transition plan, the management team should collaborate with the individual to:

- Review the first draft early in the transition period.
- Conduct at least one to two checkpoints to assess progress.
- Complete a formal review at the end of the transition period.

<u>**Note**</u>: Oftentimes in companies, the actual work a person does may encompass more than is captured in the role description. Stronger performing individuals will take on additional tasks and activities as requested by the program management team. The information captured in this step provides the individual an opportunity to document these tasks and provide candid feedback on what worked well in the role and what could be improved. This is very valuable information for a mature program office.

2.1.5.4 Referral Email

One of the most effective ways a program office can support an affected team member is by proactively providing individualized references within the company. These references can help generate opportunities that best align with the individual's skills and experience. The management team should be familiar enough with the team members' strengths to highlight their high-value skills and support their professional growth by providing candid, constructive feedback.

A referral email should include:

- A concise subject line
- A clear call to action, request, or question
- The background and purpose of the email
- Key highlights of the team member's skills and achievements
- Points of contact for follow-up
- An attachment or link to the team member's resume or profile

Below is an example internal referral email for Mr. John Doe, a project scheduler who reports to Ms. Smith:

Subject: Project Scheduler Resource Available

Content:

Project GOAL was successfully implemented and we are currently looking for roles for our talented staff who will be losing coverage in six weeks. Please reach out to our management team if you have any positions that are suitable for John Doe.

Mr. Doe performed exemplary in his role as project scheduler where he helped coordinate the schedule across five teams, ensuring compliance with our standard schedule.

You can contact his supervisor, Ms. Smith, directly, or reach out to anyone on our management team if you need additional background. Feel free to reach out to Mr. Doe directly to discuss potential opportunities and determine if he is suitable for your team.

His resume is attached for reference.

3 BUSINESS DEVELOPMENT

The objective of the Business Development section is to focus on optimizing the process of generating leads and turning those leads into value-added revenue streams. Large contractor organizations begin with the end in mind, creating an environment that supports a continuous learning environment with clear standards and procedures that will be scalable to grow and expand with the program.

Primary Goals:

- Lead/Prospect Management
- Optimizing team size for opportunity capture
- Response Standards
- Measuring Performance

Business development refers to the search, preparation, and capture of new work for a company or program. The approach differs for companies offering services, versus those selling products. However, both rely on gaining customer trust and demonstrating the value of their offerings. For service-based companies, trust is best built over time, and large service organizations invest significant resources in developing long-term relationships with their customers. Smaller organizations, on the other hand, often rely on government set-aside programs to help them sell their products and services. Small businesses should take advantage of opportunities provided by the Small Business Administration to increase their chances of success in both commercial and government contract capture.

Any team member working with customers should continuously develop their competencies in building trust. Numerous books and training resources offer further insights into trust-building strategies. In *The Trusted Advisor* (Maister, Green, & Galford, 2001), the authors highlight the key to professional success: the ability to earn the trust and confidence of clients. They outline a step-by-step process for building trust-based relationships, which includes five key steps: engage, listen, frame, envision, and commitment. One of the hallmarks of high-performance professionals is their ability to be active listeners.

Many companies specialize in providing training and services to help organizations improve their business development and capture strategies. A smaller organization may struggle with securing new programs due to limited staff, while a larger organization may focus on capturing high-value projects that align with its core capabilities. Understanding the nuances of business capture will help you determine the best approach for securing new opportunities. Once this approach is established, your company can use the guidance in this section to develop the necessary infrastructure and methods best suited to its needs.

3.1 Understanding of Opportunity and SWOT Analysis

Organizations must determine how best to position their products and services in the market to ensure sustainability and long-term viability. This requires a deep understanding of the market and how their offerings serve that market. The two most critical aspects of business development are:

1. Understanding the opportunity, evaluating potential contracts or projects to determine their feasibility and alignment with company goals.
2. Understanding how your organization's core competencies and strategic roadmap align with the opportunity, ensuring that the business pursuit is in line with long-term objectives.

3.1.1 Opportunity Review/Capture

Not all opportunities are created equal, and understanding the scope and minimum requirements of any potential contract is essential. In government contracting, there are well-established processes for soliciting proposals. Typically, a government agency will begin by issuing a Request for Questions (RFQ), Request for Comments (RFC), or Request for Information (RFI). These early-stage solicitations help agencies develop well-structured proposal requests.

It is in the best interest of any company intending to bid on a proposal to engage in these early stages for two key reasons:

1. Influence and Insight: Participation allows companies to help shape the final proposal requirements before the formal Request

for Proposal (RFP) is issued. It also provides critical insight into project scope and minimum requirements.

2. Visibility and Relationship Building: Engaging in these early stages helps establish awareness with stakeholders, some of whom may be part of the selection committee.

Once the government agency finalizes the requirements, it releases the official RFP. During this phase, there is typically a short period during which potential bidders can submit questions. This window is crucial, as all submitted questions and the agency's official responses are shared with all competing vendors. The types of questions asked, and how they are answered, can provide valuable insights into competitor strategies.

Before deciding to respond to an RFP, companies should ensure they have captured and clearly understood at least nine key pieces of information.

1. **Background and Scope**: The background and scope of the opportunity is usually one of the first sections of the RFP.
2. **Category**: If your organization has multiple divisions or core competencies, you may want to have a standard list of categories for opportunities. Opportunities may also be categorized by size or industry.
3. **New Opportunity Indicator**: Is the opportunity a new one or is it a recompete or replacement product or service? If it is a recompete, it will be important to understand which company is currently providing the product or service. Understanding your competitive landscape will always include the incumbent as they have a competitive advantage; in particular, if they work onsite or with the customer on a daily basis.

4. **Contract Type**: What is the contract type? Is the contract a firm fixed price (FFP), cost plus fixed price (CPFP), or time and material (T&M)? Understanding the various types will be required to complete the pricing portion of the response.

5. **Limitations:** Are there any known limitations contractually? Many government organizations now require a small business set aside to level the playing field a bit. The set aside is usually measured as a percentage of work that must be performed by a small business. Understanding some of these constraints allows you to determine the most suitable partners to include in the response.

6. **Period of Performance**: How long is the contract period and is it fixed or are there optional renewable periods? Many government solicitations have a base period and then option year periods. So, a six-year contract would consist of one base year and five option years which then allows the organization to determine annually if they are going to continue the contract or put it back out to bid. Unless an incumbent is really underperforming, it is safe to assume that the winner will maintain the contract for the full contract period. The amount of overhead and risk in turning over midstream is usually too disruptive and costly for the organization submitting the RFP.

7. **Value**: What is the expected value to your company? Your company may have profit goals and thresholds but that is not the only value that should be considered. Strategically, there may be other opportunities in working with a customer. Maybe it is working in a new industry or providing a new product that is not well known.

8. **Submission dates**: What are the key dates for response? For mature response cycles this may include the dates for the RFI, RFC, or RFQ; proposal due date; oral presentation date;

expected award date; contract start date, etc. The more information you have, the better it is to plan for both the capture activity and the execution of the work.

9. **Key Personnel**: Are there essential roles indicated in the contract? The essential roles are usually roles that require official customer notification if changed and provide key strategic positions for the project. One of the most important evaluation criteria is how strong the proposed key resources are that will be assigned to the project. The proposal team should ensure they have contingent offers for the individuals identified for these key roles to be in a better position to win.

Additional consideration is optimizing a good internal capture support team. Clearly identifying the members of the team, pulling the response together, and their roles as the primary facilitator for each proposal is important. There should be no fewer than three separate people as part of the team, the authors, the reviewers, and the quality assurance person. On larger pursuits, there may be additional roles assumed for pricing, capture facilitation, business development, technical writing, subject matter experts to act as both content authors and content reviewers, production staff, and an executive review committee for each capture gates for more formal responses.

3.1.2 Strengths, Weaknesses, Opportunities, and Threats (SWOT) Analysis

One of the better techniques when starting to look for new business is to perform a SWOT Analysis. This strategic technique is used to help organizations focus on both internal strengths and weaknesses and external opportunities and threats. Having a clear understanding of these factors allows the organization to target new business that best aligns with core values and increases the success profile.

Strengths are also known as core competencies. They include proprietary technology, skills, resources, market position, patents, and so on. As you assess your strengths, you should consider your long-term strategy. Questions to consider are:

- What are your core strengths now? What should your customers know that you are good at?
- Are your customers aware of these strengths? If not, what is the communication/marketing plan to address this?

A **weakness** is a condition within a company that can lead to poor performance. Questions include:

- Are there other strengths the sponsoring organization expect your company to have for this specific opportunity? If so, do you have time to develop and communicate these as strengths?

Opportunities are current or future conditions in the environment that a company might be able to turn to its advantage. Considering external opportunities, the organization needs to determine which opportunities are best suited to explore. Larger organizations typically set thresholds due to the amount of time commitment required to increase win rates of opportunities. As opportunities are explored, a few questions to consider include:

- Does the opportunity align with and support the strategic vision for the organization?

- What benefits specifically does the organization receive from the opportunity?
- What are the impacts if we win or lose the opportunity?
- How many resources can we afford to expend to increase our chances of winning?
- Does your organization have any existing relationships or are there opportunities to forge new relationships with the sponsoring organization?
- Do you need partners to increase your chance of success? If so, is your organization going to be the prime, sub, or equal partner?

Threats are current or future conditions in the environment that might harm a company.

- Who are your likely competitors for the specific opportunity?
- Is there an incumbent or is this new work?
- Are there any other niche vendors that competitors may partner with for strategic advantage?
- Are there other strategic partners that will allow your team to maintain a competitive advantage for this proposal?
- Are there any limitations or constraints?

It is important to note that as an organization matures in this area, alternative techniques may become more suitable. One such technique is Porter's Five Forces Analysis. This framework analyzes the competitive landscape of a business, drawing from industrial organization (IO) economics to identify five forces that determine an industry's competitive intensity and overall attractiveness in terms of profitability. Originally introduced in the Harvard Business Review's

March-April 1979 article, *How Competitive Forces Shape Strategy* (Porter, 1979), this framework has since become a cornerstone of elite management programs.

3.2 Methods

Many organizations have developed entire industries around training professionals in the most effective business capture methods. One of the most prominent is **Shipley and Associates,** which provides templates and training for large organizations to enhance their business development efforts. According to Shipley, *"Shipley certification adds value to your experience and education and helps improve your position in the job market"* (Home, 2024). Having a certified resource on your high-performing business development team can serve as a key differentiator. The same methodologies can be adapted into a scaled-down version for smaller companies or individuals seeking to secure government contracts.

At the heart of the method is a formal review process. To ensure an efficient and effective response, a structured approach should be followed:

☐ **Facilitator:** One person should be designated to oversee and guide the process.

☐ **Authors & Reviewers:** Specific team members should be assigned to each phase of the proposal.

☐ **Compliance Reviewer:** A separate individual should review instructions and evaluation criteria to ensure the response fully addresses the questions and aligns with the scoring criteria.

> **Note**: Capture Management has become an industry, and if you do not have an experienced person on your team, it is highly recommended that you seek out a Certified Capture Manager (CCM™) to increase your chances of success.

There are **six common color review gates** that a **capture manager** should be familiar with to facilitate better communication throughout the proposal process. Numerous books provide more in-depth insights into this topic. The **Capture Management Competency Table** (Table 3: Capture Manager Review Gates) below summarizes each review phase and its associated goals.

Review Gate	Description
Blue Team Review	**Conduct a Blue Team Review to assess the capture plan and strategy.** A Blue Team review ensures that the basic **outline of the proposal is correct and complete** and that a writer is assigned to each section. The Blue Team review also identifies gaps in information, data, designated SMEs, discriminators, and win themes.
Black Hat Review	**Conduct a Black Hat review to predict competitor's likely solutions and strategies.** The Black Hat Review in some organizations takes the Opportunities and Threats (outside forces) to another level. Understanding the competitive landscape is an important part of how to position and differentiate your response and solution against those who are most likely to be bidding for the same work.

Review Gate	Description
Pink Team Review	**Conduct a Pink Team Review to verify compliance and execution of your win strategy.** Some organizations describe the goal of a Pink Team document as **65-70 percent complete**. A more useful definition is to describe the expectation that the document addresses with content, approach, or intent in every section of the proposal. The Pink Team document adds narrative and detail to the outline developed for the Blue Team. This document contains graphics and tables. Emphasis is on content, not on form, style, or grammatical perfection. At the same time, the document should begin to incorporate consistent ways of spelling, acronym use, common quantity references, and common ways of referring to discriminators.
Red Team Review	**Conduct a Red Team Review to predict how your proposal will be scored and make improvements to it.** For a Red Team Review, the proposal document should be nearly complete. Some organizations quantify it by saying **85 to 90 percent**. All the changes, additions, and corrections identified at Pink Team Review should be incorporated in this document. The document should have final or near final graphics, tables, data, and resumes. All sections should have complete narratives and the documents should be formatted the same as it will be submitted without going through a formal desk top publishing edit. In reference to this last point, depending on the review process, the document may or may not retain the full RFP references but should retain RFP paragraph references after the heading titles.
Green Team Review	**Conduct a Green Team Review to review and approve pricing.** The Green Team Review is also referred to as the pricing review. This review may occur at any time but often it is at or after the Red Team but before the Gold Team review. Normally, if the proposal manager or writing team are contracted support, they will not be involved in pricing development or review. The review documents are the pricing information in the formats prescribed by the RFP.

Review Gate	Description
Gold Team Review	**Conduct a Gold Team Review to confirm your proposal incorporates necessary changes from Red and Green Teams, and is ready for the Proposal Submittal Decision.** The Gold Review document is pre-submission quality. All sections, all information and all graphics are complete, and fully compliant. It is formatted and ready for submission.
White Hat Review	**Conduct a White Hat Review to record lessons learned and make process improvements.** The White Hat Review, similar to a retrospective, is the way the organization capture management team implements a continuous improvement model. Lessons learned during the capture process for each proposal are documented, and specific action items and action item owners are identified to implement in future capture opportunities.

Table 3 Capture Manager Review Gates

3.3 Infrastructure Setup

The project manager's role in business development is often referred to as the capture manager. In this capacity, the manager is responsible for assigning the best resources to complete the proposal for submission, communicating the response capture process and schedule, providing the necessary templates for efficiency, providing a collaboration site for the draft submission of sections of the document, and ensuring a high-quality submission via a thorough review process. The size of the team may depend on the size of the company, the revenue potential of the opportunity, and the availability of resources.

The first step after the company has identified and approved an opportunity for capture is to set up the collaboration site. Access should be granted to each capture team member once the Request for Information (RFI), Request for Proposal (RFP), or Source System Notice (SSN) is published. The main milestone dates should be captured and highlighted in a separate area, including critical dates such as the deadline for questions and the proposal submission date and time. These milestones will serve as anchor dates for the proposal work schedule. The directory structure should be created to organize necessary work folders/areas for each review cycle and to act as a reference directory, including supporting items such as previous relevant proposals and work products.

Once the collaboration site is created and the originating supporting artifacts are loaded, it is time to create a few standard documents to enhance efficiency. The first document should be the proposal checklist, which serves two purposes. The first is to provide expectations for writers and reviewers regarding all requirements that must be addressed, ensuring a structured writer's delivery template.

The second purpose is to serve as a key component of the Gold Review. As one of the final review stages, the checklist enables the appropriate team members to verify that all requirements have been met. Too often, even within large capture teams, a submission can be disqualified due to a simple requirement omission.

Standardization and clear communication of processes are recurring themes throughout this guide. In new business development capture, the need for a company to have a well-vetted process is particularly evident, as these types of projects typically operate on a fixed timeline of 2-8 weeks from proposal posting to submission. Since multiple individuals are often involved in writing and reviewing, it is crucial that a draft template for the response is created. This document should contain several important components that are reviewed early in the process as part of the Blue Review. The template should include an outline of the response formatted per requirements. It is beneficial to provide a section overview so that writers have a reference readily available for each portion, and reviewers can clearly understand the questions being addressed. More mature processes also incorporate 'win themes', which are differentiators that set the company apart from its competitors and should be integrated throughout the proposal. Using the template requires adherence to engagement rules, such as ensuring that any reused content from prior proposals is formatted according to the target document. The cleaner the document is after each writer's contributions, the more effective the quality review process will be leading up to submission.

It is important to identify the best person on the team for each role. Equally important is establishing a standardized method for collecting and communicating reviewer feedback so that it can be quickly consolidated and shared with writers. Implementing a structured mechanism or tool for individuals to provide constructive feedback to the proposal writing team is a simple yet effective way to manage time constraints and enhance communication between reviewers and writers. The facilitation tool should capture both summary information for high-level review discussions and detailed feedback for the writers. Review guidelines should be incorporated into the instructions or embedded within the template to ensure each reviewer adheres to a consistent approach. These guidelines may include references to win themes, proposal compliance, writing clarity, and the use of graphics. Providing structured guidance on recommended changes helps reviewers offer actionable suggestions, allowing writers to quickly implement necessary updates to produce a high-quality proposal.

Any organization with a strong business development focus must establish a process for tracking key performance metrics for each response. Defining these metrics early in the process enables the organization to assess the effectiveness of its business development efforts and identify areas for improvement. Simple metrics, such as tracking win/loss outcomes, to more complex analyses like evaluating pricing strategies, can provide valuable insights for refining capture strategies. It may be as straightforward as observing pricing trends to determine an optimal target price for a specific customer type. Alternatively, an organization may recognize that a particular team has a consistently high win rate and leverage its expertise to train other teams with lower success rates. By capturing and analyzing the right metrics, organizations can enhance their strategy and optimize the time spent securing new business.

Understanding the principles of effective capture management is essential for organizations. These principles improve communication, set clear expectations, and ensure customers receive the information they need to make informed decisions. A key measure of an organization's capture success is its win rate. Establishing appropriate thresholds in this area allows the company to determine whether additional time and effort are needed to enhance proposals and improve team performance in securing work.

Setting up the infrastructure provides a foundation for maturing the response process and developing reusable proposal components over time. As with any process, the more frequently an organization engages in capture activities, the more experienced and efficient it becomes. Continuous refinement of processes enables organizations to streamline their efforts, reduce waste time, and increase the probability of success. The next focus should be on the process of preparing the response.

3.4 Preparing the Response - Approach

This section provides a brief overview of the minimum requirements for setting up a team to provide a high-quality response in an efficient manner. There are three primary roles that should be targeted. The three roles are:

1. The response content authors,
2. The response content reviewers, and
3. Quality assurance.

3.4.1 Response Content Authors

The response content authors are usually individuals familiar with the work outlined in the proposal's scope. Their primary focus is to develop the core content of the response. The writing process can be simplified into the following steps:

1. **Scan all relevant customer material:** Relevant customer material may include any related SSN, RFI, RFQ, RFC, the official RFP, questions and answers on the official RFP (if available), and any amendments to the original RFP.

2. **Have any instructions, conditions, and notices readily available:** If the author is using a computer with multiple screens, these documents should be accessible on one of the screens. Alternatively, printing them out for easy reference can be helpful during content development.

3. **Ensure the evaluation criteria is readily available:** The evaluation criteria document is the most critical reference for the content author. It provides a rubric that outlines how evaluators will assess the responses. Therefore, the author

should write clearly and directly to these criteria and review them before and after drafting each section.

4. **Reread the Questions and Answers:** The questions and answers should be reviewed for any relevant details that apply to the section being written. Any pertinent questions should be reread to assess changes made to the original RFP and gain insights that may be useful in completing the response.

5. **Review and understand your 'Win Themes':** Win themes vary by company and define what differentiates the company from competitors. Before beginning to write, the author should confirm they have a clear understanding of these themes. The more authors who are contributing to the response, the more crucial this becomes. One of the capture manager's responsibilities is to work with the team to define these themes early in the process, ensuring consistent messaging and reducing the need for extensive rewrites later.

6. **Read assigned sections:** Each assigned section should be worked on individually, following steps 3-6. Once the author has reviewed the supporting documents, they should focus on their assigned section and its corresponding evaluation criteria.

7. **Write each section according to the instructions and evaluation criteria:** Writing each section may require multiple iterations. The author should start by outlining key points that need to be addressed. Then, they should develop the response based on the understanding gained from steps 3-6.

8. **Verify alignment with evaluation criteria:** The author should review the draft to ensure it meets the evaluation criteria. This second iteration helps improve clarity and alignment with requirements, enabling reviewers to provide more constructive feedback and leading to a higher-quality final response.

9. **Repeat for other assigned sections**

10. **Conduct a personal quality check:** It is beneficial to step away from the writing for a short break—stretch, go for a walk,

or engage in another activity. Returning with a fresh perspective allows the author to reread the content and verify flow, messaging, spelling, and grammar. This ensures that reviewers can focus on the content's quality rather than correcting minor errors.

11. **Submit the section for review:** Once the section is submitted, take another break. Engage in other activities, relax, or exercise. Allowing for mental recovery is just as important as the time spent writing.

3.4.2 Response Content Reviewer

The response content reviewers are usually composed of individuals who are familiar with the work that is in the scope of the proposal and have experience responding to similar types of proposals. The focus for the reviewer is to review the content and confirm the section is written clearly and meets the intent of the RFP. The process for the reviewer can be simplified into the following steps:

1. **Read/Scan all relevant customer material**: Relevant customer material may include any related SSN, RFI, RFQ, RFC, the official RFP, questions and answers on the official RFP, if available, and any changes to the original RFP.
2. **Have any instructions, conditions, notices readily available:** If the reviewer is using a computer with multiple screens, these documents can be readily available on one of the screens. Alternatively, printing these out and having them available for the review should be helpful during the content review cycle.
3. **Have Evaluation Criteria section readily available:** The evaluation criteria is the most important reference document for the content author and reviewer. It provides a rubric that

specifies what the evaluators are using to judge the responses. As a result, the author's role is to answer specifically and clearly to these criteria and, as a reviewer, the criteria for each section should be confirmed while reviewing the content.

4. **Reread the Questions and Answers:** The questions and answers should be scanned for any relevant items that pertain to the section being written. Any pertinent questions should be reread and assess changes made to the original RFP to have a better understanding of any insights that may be pertinent to completing the review of the response in the section.

5. **Read and understand your 'Win Themes':** Win Themes are different for every company. They tell the customer what sets you apart from your competitors. Before the review begins, the reviewer should confirm that they understand the win themes. The larger the number of contributing authors and reviewers to the response content writing, the more important this becomes. One of the responsibilities of the capture manager is to work with the team on defining these themes so that everyone has clear directions on messaging early in the process. This saves much time on rewriting later in the response process.

6. **Read assign section being reviewed:** Assigned sections should be worked on one at a time following steps 3-6. Once the reviewer has read the supporting documents, focus should be placed on the section and the associated evaluation criteria.

7. **Review the section according to instructions and evaluation criteria:** Does the author provide clear and concise responses to the evaluation criteria? Does the section flow? Are there any recommendations for improvement? The reviewer should include inline comments with any suggested changes. On small teams, using a documentation application that includes the 'Track Changes' feature is one approach that can be used

effectively. The inline comment approach is recommended and becomes more important as your capture team size grows as it facilitates the consolidation of multiple authors and reviewers to be more efficient. The more efficient the process, the more time can be spent on high quality content development.

8. **Confirm writing against evaluation criteria:** The reviewer should confirm that it meets the evaluation criteria. This second iteration will lead to better review and recovery by the authors. Again, inline comments should be used to provide specific feedback and guidance to help facilitate the author's recovery effort.

9. **Repeat for other assigned sections**

10. **Re-read and summarize main feedback**

As you review the section, feedback may be varied by importance. It is important to succinctly summarize the main suggestions to make the most improvement with the often-limited time for this exercise. Scoring each section helps to identify which sections require more focus and priority for recovery and is helpful with time constraints. The scoring should be clearly defined so that multiple reviewers are standardized in their response and feedback. Scoring techniques could be a simple rating of 1 to 5, clearly defined, and color coded. An example of a simplified approach is included in the table below:

Score	Color	Description
1	Red (Unacceptable)	The section is not acceptable. This may be due to: • Section does not address most evaluation criteria • Section is missing a lot of important requirements • Section is ambiguous and does not clearly answer the response • All relevant Win Themes are not included
2		The section is marginally acceptable. • Section does not address all evaluation criteria • Section is missing important requirements • Section is ambiguous and does not clearly answer the response • All relevant Win Themes are not included
3	Green (Acceptable)	The section is acceptable. Improvements are available for higher quality. • Section addresses evaluation criteria • Section has no missing requirements • Section is understandable • At least one relevant win theme is included
4	Purple (Good)	The section is acceptable. Improvements are available for higher quality. • Section clearly addresses evaluation criteria • Section has no missing requirements • Section is clear and concise • **Most relevant winning themes are included** • **Minor comments for recovery**
5	Blue (Delivery Ready)	The section is acceptable. • Section clearly addresses evaluation criteria • Section has no missing requirements • Section is clear and concise • **All the relevant win themes are included** • **No comments for recovery**

Table 4 Reviewer Scorecard Summary

Comments should also be captured focused on the SWOT analysis (see **Section 2.1.2**) or by capturing the section strengths and weaknesses along with any risks or deficiencies. The clearer the review, the easier it will be for the author to recover.

11. **Submit section for recovery:** On larger more mature capture teams, a formal meeting between the authors and reviewers is scheduled. In this meeting, the reviewers can summarize their findings, suggest improvements and answer any questions the authors have on the reviewers' comments. Getting this objective, outside perspective allows the author to review and modify, thereby creating a higher quality deliverable.

3.4.3 Response Quality Assurance

The quality assurance role in a response has two main goals. The first goal is to ensure that all evaluation criteria are met. Some RFPs explicitly require the inclusion of evaluation criteria in the section responses to assess the evaluation team. If a weighting is provided, it is beneficial to include it in a table early in the process, as this may help authors prioritize sections if there are time constraints.

The second purpose of the quality assurance role is to confirm that all instructions are followed. This includes requirements such as format specifications, page limits, and specific information requested in the RFP. This role acts as a traffic controller, ensuring that all response guidelines are adhered to and that the submission is complete.

Note: The importance of this Quality Assurance role cannot be emphasized enough. Much good work and effort that go into creating a high-quality deliverable can be negated due to a missed requirement that causes an automatic disqualification.

3.4.4 Other Capture Roles

Additional important capture team roles that support the business development effort may include, but are not limited to:

- **Technical Writer:** A skilled technical writer reviews the entire response and ensures a unified voice across sections written by multiple authors. This role covers everything from basic corrections like spelling and grammar to more advanced writing techniques, such as integrating the win themes into a compelling narrative. Clear and concise communication of how the product or service will be delivered—and why this company is the best choice—is essential for a winning bid.
- **Production:** The production team focuses on polishing and finalizing the proposal response. Their responsibilities include ensuring clear titles, properly formatted graphics, embedded images, and tables that align with the proposal's requirements. If the requestor does not specify a minimum text size for graphics, it is important to establish internal standards to ensure readability. Well-designed graphics can be a game-changer, particularly for proposals with strict page limits.
- **Pricing Team:** Experience and market awareness play a crucial role in pricing strategy. Many factors influence pricing, including the requestor's expected budget, market rates for

applicable resources, resource availability, and the company's revenue and pricing threshold expectations.

- **Business Development:** The business development team is primarily responsible for building and maintaining relationships with the customer. Meetings are typically held throughout the year to gain insights into the customer's strategic direction, pain points, and upcoming initiatives that align with the division's goals. Many larger organizations will not pursue opportunities unless this groundwork has been laid before an RFP is issued. Mature organizations recognize the value of relationship-building in securing new business and dedicate significant time and resources to cultivating these connections. This approach often provides them with a competitive advantage over companies who first engage with the customer only when the proposal is released.

4 RECRUITMENT MANAGEMENT

The objective of the Recruiting section is to prepare for hiring high potential candidates quickly.

Primary Goals:

- Recruiting Team Makeup
- Schedule Alignment
- Standard Operating Procedures
- Measuring Recruiting Efforts

Recruiting is one component of a program where process improvement areas can be quickly identified and matured. A strong and competent program office team is critical to achieving success. Senior members of the program office must set the tone for both the office and the entire program team. The program support team should have enough experience to execute the processes defined and implemented by the program. Additionally, a junior but motivated resource can thrive and add significant value to the team. The program office can guide the career paths of support team members based on their areas of interest.

The recruiting process is easier to optimize for smaller projects, and the value of process improvement techniques outlined in this section increases with the size of the program. The four perspectives considered in this area of focus include process improvement for the interviewer, the recruiting team, the prospects, and the program management support role.

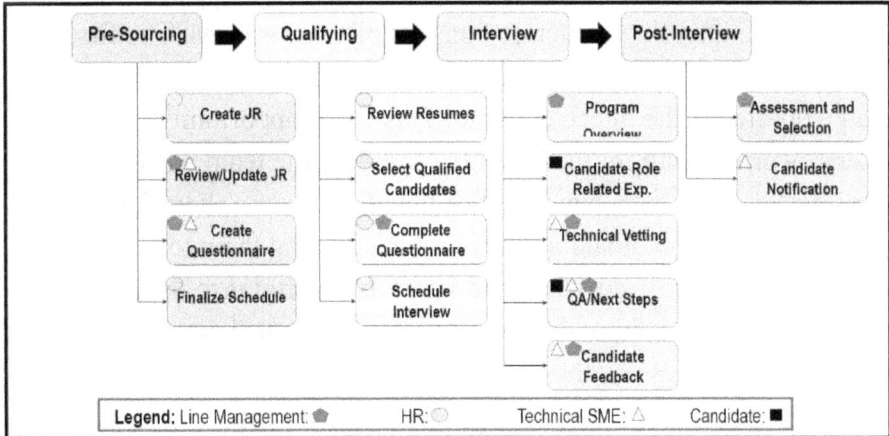

Figure 3.0: Recruiting Process

4.1 The Prospect Interview

Conducting an interview effectively is both an art and a science. Pairing experienced interviewers with team members who are participating based on their position or title allows less-experienced resources to develop the skills required to make the interview process a positive experience for all stakeholders—including the prospect, the interviewers, and the project support team.

The individuals conducting the interviews are usually members of the management team leading various program teams, including technical managers. A best practice is to have at least two individuals conducting the interview. The team should timebox interviews to at least 30-minute intervals. Each interview should begin with an introduction, including thanking the candidate for their time and providing a brief overview of how the interview will be conducted.

An effective interview structure consists of four sections. The first section is the welcome and introduction, which should take no more than a quarter of the interview time. This is the opportunity for the interviewer to highlight the benefits of joining the team and explain the importance of the role. Next, half of the allotted time should be dedicated to the candidate's background, allowing them to communicate their experience and value as it pertains to the role. During this portion, interviewers can ask specific technical and soft skill questions to assess the candidate's fitness for the program.

The final portion of the meeting should be set aside for the candidate to ask questions. One of the interviewers should also explain the next steps in the process. This approach allows the technical interviewer to minimize their time in the interview while still ensuring the prospect receives the engagement necessary to secure their interest. Since technical managers are primarily focused on delivery, their questions tend to focus on assessing whether the candidate meets the minimum skill requirements. Therefore, it is recommended that each interview includes both a subject matter expert and a program management team member who can assess soft skills and promote the program. The acceptance rate of well-suited prospects is proportional to the time invested in this portion of the interview and should be monitored as part of ongoing program improvement.

4.1.1 Interviewers

Each interviewer plays a crucial role in conducting a successful interview. The first interviewer is responsible for selling the program. As programs expand, it becomes less effective to have only one or two individuals conducting the interview with a sole focus on assessing the prospect's value. A common challenge is that technical managers are often too busy to fully participate in interviews. One solution is to pair interviewers—one from the management team and one from project management leadership. This approach allows the project management office to introduce and close the interview while the technical interviewer focuses on assessing the candidate's technical skills.

4.1.2 Interviewer Availability/Schedule

The project support team member assigned to recruiting must first determine interviewer availability. This can be accomplished by independently meeting with each interviewer for a brief discussion to identify preferred time slots. Using the program's recurring meeting calendar (see **Section 4.2.1**), the support team member can propose weekly interview time slots.

This approach is particularly useful when there is a surge in hiring needs, requiring additional time from the management team. Once confirmed, a recurring interview meeting can be scheduled as a placeholder. The focus then shifts to filling the available slots with potential prospects. These scheduled slots can be incorporated into the Recruitment Team Availability Schedule. If a particular slot remains unfilled, the support team member can use that time to review and prioritize candidates. Another approach is to send the list of candidates to the team leader for prioritization. Prioritizing candidates ensures efficient use of time. If an ideal candidate is identified early, the remaining prospects can be considered for other openings or notified that the position has been filled, saving time for all involved.

4.1.3 Interviewer Pre-qualification Questionnaire

Not all programs require a pre-qualification questionnaire, but it is beneficial for larger programs. The need for such a questionnaire often becomes apparent during a recruiting cycle if multiple interviewees are not selected for hire. Interviewers can prepare this document to filter out unsuitable candidates early in the process, preserving interview slots for those who meet a predetermined threshold.

Experienced technical interviewers often have a standard set of questions they use to assess candidates. The program management support team can identify repeated questions during initial interviews and work with interviewers to document a standard set of pre-qualification questions. More importantly, the interview team can collaborate to develop a pre-qualification questionnaire for the recruiting team to use when initially screening prospects.

Tracking and analyzing the Prospect Acceptance Rate metric can help the recruiting and project management support teams refine the pre-qualification process, leading to continuous improvements.

4.1.4 Interviewer Assessments

All programs should have a standardized approach for fairly and consistently evaluating a candidate's skills and overall fit within the team. The assessment should be flexible enough to capture both soft and technical skills. A ranking scale from one to ten can be used to rate each skill type.

Understanding how each interviewer ranks the candidate can lead to better questions from the management team and provide a clearer view of each candidate's strengths and weaknesses. A separate section should capture general comments and include a specific 'go/no-go' recommendation.

The project manager is typically responsible for consolidating feedback from all interviewers and relaying it to the program support team for further action. If time allows at the end of the interview, feedback can be quickly provided via email using a structured template to score candidates. This approach reduces the interviewer's workload while ensuring valuable input is gathered for the project and program management teams.

The objective is to optimize the time spent on interviews, enabling a swift and informed hiring decision. The goal is to onboard the best candidate as efficiently as possible, thereby alleviating the workload stress that necessitated the hire in the first place.

4.2 Recruiting Team

On larger programs with the advantage of a dedicated recruiting team, process improvements can often be quickly identified, particularly in communication management. It is crucial that the management team remains engaged rather than disengaging, ensuring an efficient model where the project team can focus on customer delivery. The project management support team members play a key role in coordinating with the recruiting team to optimize the most effective collaboration methods, minimizing effort while ensuring that recruiting sessions are valuable for both teams.

The primary goals of the initial meeting between the program and recruiting teams should be to clearly define the program's specific hiring needs and determine the most efficient way to interact with the recruiting team. This includes identifying any internal status reporting required for upper management and establishing best practices for initiative success. For instance, setting a consistent interview schedule—such as a designated day and time each week—based on stakeholder availability can significantly improve efficiency.

4.2.1 Weekly Touchpoint

A weekly touchpoint should be established to review all open job requisitions. In larger programs, any funded but unfilled position is typically considered lost opportunity revenue for the company. A common goal is to fill all open positions within 30 days.

With this key performance metric in mind, the project management support representative should collaborate with the recruiting team to set up a dedicated weekly meeting focused on open job requisitions. Each unfilled position should be discussed, and its status documented. To maximize effectiveness, this meeting should be scheduled no later than one day before the internal program management status meetings with the company's upper management team.

4.2.2 Recruiting Pre-qualification Questionnaire

A recruiting team's primary function is to sell the program and identify qualified candidates. The pre-qualification criteria typically include factors such as citizenship, education, years of experience in a specific technology, and other contractual requirements defined by the recruiter. Establishing a clear understanding of these criteria is a critical first step in aligning the project and recruiting teams.

The Recruiting Pre-Qualification Questionnaire can also be expanded to incorporate elements from the Interviewer Pre-Qualification Questionnaire to enhance candidate selection. A strong recruiter can better assess the program's needs when the interviewers provide well-structured pre-qualification questions. This alignment ensures that only the most suitable candidates move forward in the process, improving efficiency and success rates.

4.2.3 Prospect/Candidates

Numerous studies highlight the key factors that influence candidates' decisions to accept a position. High-performing team members are often driven by the program's mission, the enthusiasm of the team, and leadership commitment to their professional growth. A recruiting process that prioritizes the candidate's experience yields significant benefits, including a reduced number of interviews and a higher acceptance rate among qualified candidates.

The interview and scheduling process serves as a candidate's first impression of the organization and its leadership. When executed correctly, the initial stages of recruitment can have a lasting impact on a prospect's perception of the company. The project support team members must be personable and actively support candidates throughout the entire process—from the initial call through the interview and final hiring decision.

While the recruiter remains the primary point of contact and holds the final hiring authority, additional project-level support for candidates enhances program efficiency. Providing a seamless and engaging experience from the outset fosters a positive impression, ultimately contributing to a successful recruitment process.

5 INTEGRATION MANAGEMENT

The objective of Integration Management is to effectively combine two 'different and complementary models' (S, 2010) to better produce viable solutions that align with the customer's vision and goal of the program.

Primary Goals:

- Manage Stakeholder Expectations
- Develop Consistent Code
- Optimize Standard Procedures to Minimize Technical Debt
- Risk and Quality Management and Mitigation
- Measuring Outcomes

PMI's Global Congresses provides practitioners with an opportunity to learn more about important concepts in the field. In 2010, Stefano Setti provided one such example that sums up integration management as combining two different work models, project and process management.

"The mission of the project is to reach the goal and stop."

"The mission of the process is to sustain the business in a continuative manner."

I interpret this as a good way to distinguish between a program and a project. The program is higher-level and defined as an ever-evolving continuous improvement in sustaining an organization. The focus of the program is evolving through the project management ideology, continuously advancing business through distinct projects that provide specific value to the organization. In this chapter, the purpose is to focus on elements of the process that we can use to minimize the individual project risk.

5.1 Customer Management

The objective of the Customer Management section is to identify the primary stakeholders of the services and solutions being provided and to understand their standard operating procedures. This knowledge enables project communication to be aligned in the most efficient manner with stakeholders.

Primary Goals:

- Stakeholder Mapping
- Schedule Alignment
- Standard Operating Procedures
- Measuring Value of Solutions

Managing customer expectations is one of the primary responsibilities of the program manager and their management team. Organizations typically have high expectations for the consultants they employ, and business needs often evolve, requiring flexibility in order to successfully partner with the organization and achieve its mission. At the same time, your company must effectively plan and allocate resources to provide the best value to the customer.

Many organizations struggle with scope management, a key determinant of project success. As a project manager, addressing scope challenges effectively can mitigate stress on the program and position the program office as a leader in exceeding customer expectations while balancing profitability and partnership. The most effective organizations implement a formal approval process for new projects, incorporating input from the contracting team. Project managers need tools to quickly assess high-level estimates, ensuring that new initiatives align with business objectives and create measurable value.

1. Identify new opportunities and the initiative sponsor
2. Include any new initiatives to be discussed as part of the recurring meeting agenda with the customer
3. Use the meeting time to understand priorities and impact of in-flight initiatives
4. Document the reprioritized priority of in-flight initiatives for the team

Preparing for meetings with program sponsors should not be the most time-consuming activity for the program office. Instead, preparation should naturally flow from existing processes designed to move a new request from concept to implementation. The first step is to understand the scope of the new work, expected delivery dates, and insights into the initiative's purpose and value.

In larger programs, this internal process can be further matured and standardized as part of a New Project Intake Process. Ultimately, all new requests require approval from both the company and customer management teams.

To ensure efficiency, the consulting organization should complete the request and secure sponsor approval before initiating any work on the initiative. Over time, companies and their customers develop a rhythm for this approval process, often refining it as part of recurring meetings. However, teams frequently invest too much effort in preparing high-level estimates only for initiatives to be rejected. To optimize this process, an iterative approach to maturing new work approval is recommended, typically involving three key phases:

☐**Initial Evaluation:** Confirm the work meets specific criteria warranting an estimate.

☐**Preliminary Estimate:** Gather sufficient details to generate a **Rough Order of Magnitude (ROM)** for the initiative.

☐**Final Approval:** Once the initiative is approved, develop a detailed estimate and schedule.

5.1.1 New Initiative Request

New initiative requests should provide basic information which may require some preparation time to make the process efficient. Always consider the information the customer will require for making an informed decision. Background information including the business value, a summary of the project, the scope with underlying assumptions, complexity level, cost, and any known risks should be included in preparing the preliminary review. Even this preliminary initiative request requires effort on behalf of the team doing the work. A process should be created, clearly communicated, and given the appropriate level of attention throughout the life of the program. Some customers expect the details of cost and specific deliverables so understanding what drives the decision helps in developing the tool used for the new request. One challenge encountered is understanding how best to break down the work to provide estimates which may differ slightly across project types. Some organizations may use specific components like the number of screens and fields, complexity of each screen, number of extracts, complexity of each extract, number of users, and the number of installation sites of an application to determine the size and level of effort. More mature customers have initiative templates and require justification to determine which projects or initiatives will provide the highest value. It is important to communicate with standard forms and embrace known processes to improve efficiencies where appropriate. At minimum, the team should capture:

- Who is the **business sponsor,** the person who is paying or taking ownership of accepting the solution?
- What is the **formal title** of the initiative?
- When is the **request received by** date?
- What is the justification of the request including the **Description, Purpose, and Value Proposition**?
- Who is the **audience** for the solution?
- Are there any specific **prompts, filters, and constraints** required for the solution?
- Are there any **attachments** that help define the definition of 'done', including mockups or screenshots?
- Any **additional comments** regarding the request not captured above – an open-ended question.

Establishing a standard process flow from idea to solution delivery is most helpful if it includes links to the supporting systems that automate parts of the process. Standard templates can be used to capture new initiatives on the project level and similar templates can be used for project level capture for business intelligence extract, reports, and dashboard requests. *See **Section 5.6.2 for more information**.*

5.1.2 Delivery Schedule

Before starting any new initiative, a delivery schedule must be established. This ensures alignment between executive decision-making and the project execution team.

The same information used for high-level approvals should be included in a more detailed Work Breakdown Structure (WBS) for the project team. For Agile teams, a high-level roadmap should outline the specific capabilities expected for delivery throughout the project lifecycle.

The necessity for a clear roadmap grows in proportion to the complexity and scale of the program. By defining key milestones and expected deliverables early, teams can ensure that the scrum approach aligns with customer priorities, maximizing the value delivered.

5.2 Communication Management

Effective program and project management offices require strong communication skills. Leading practitioners in this field estimate that communication accounts for over 80% of a project manager's responsibilities. A strong project manager must effectively communicate expectations to the team, interpret progress and risks, and, most importantly, engage with customers and executive stakeholders. Managing various personality types is a crucial leadership skill in optimizing high-performance teams. The following steps are recommended to enhance communication and maximize the value of meetings:

1. Understand when the most important recurring client meetings are scheduled and create a recurring meeting calendar.
2. Prepare a communication packet for recurring meetings with the customer.
3. Create or update the standard communication process for capturing information required by the business community.

These three communication artifacts alone can help avoid 90% of the problems that typically hinder projects and enable effective communication with the customer.

5.2.1 Recurring Meeting Calendar

The purpose of the recurring meeting calendar is to document ongoing meetings across the program. Once established, it should not require frequent updates. Capturing this information allows for efficient team time management. While a calendar tool assists with scheduling, a structured approach ensures alignment across a growing team. With a fifteen-member project team, establishing consistent communication is easier than with a team of over a hundred members. Meeting planning should minimize conflicts through the following process:

1. **Identify the main customer meetings**

    Most organizations have a regular roster of meetings, including program, functional, and organizational updates. Documenting the most critical customer status meetings is essential. These are the meetings where the sponsor reports program progress to their leadership. Contractors may or may not be included, depending on the topics discussed. Critical customer meetings should be color-coded for visibility.

2. **Determine or adjust the program management meetings**

    Identify the most critical status meetings with the customer, program, and management team. The most crucial meeting attended by both the customer and program manager should align with the primary customer status meeting. If the customer requires materials in advance, the program manager should schedule the meeting to provide input before the sponsor's meeting. Misaligned meetings lead to inefficiencies, as customers often seek last-minute updates for internal preparations.

3. Determine the time for the other management meetings

Once the core meetings between the program management team and the customer are established, the next step is to document and revise the management team meetings to align with the information required for the meetings determined in step two. Ideally, you would want those meetings that do not require multiple disciplines to occur at the same time. For example, if each of the development teams meet by discipline of business, design, programming, and testing, these meetings should occur at the same time since they will likely be separate meetings with little overlap. During these types of meetings, the purpose is to primarily discuss standards, process improvements, and status within a competency. These meetings rarely require individuals from other disciplines unless someone from another discipline is invited to discuss a specific topic. These disciplined focused meetings are important meetings for team continuity, and to understand how the disciplines are developing the best standards and processes to support the program. Similarly, there will be cross-discipline meetings. These occur most frequently on Agile programs where each scrum team is a self-managing team and attends scrum meetings. Scrum meetings usually occur early in the morning and may be followed by a scrum of scrums depending on the size of the program. High performing teams keep these meetings true to the Agile manifesto of 15 minutes. One recommendation is to leave 15 minutes between the daily scrum and the scrum of scrums. This allows some time for follow up between the scrum masters as they prepare for the scrum of scrums meeting. Using this approach, one hour should be reserved every morning between 8-10 for these critical meetings. One advantage of this approach is that the entire team has a very specific action that will occur during the day to achieve their goals.

4. Communicate the true meeting availability to the team

Once the customer internal, the program, the team lead, and the scrum meetings are established, the entire team can identify recurring meeting times with other customers and project stakeholders. Capturing the availability of the core management team allows the program management office to create other internal meetings such as the quarterly all hands meeting, internal strategic discussions, recruiting, or onboarding sessions.

This recurring meeting calendar is a living document and it typically does not change often as it starts to form a rhythm with the team. Individuals will look at an outlook calendar, schedule an event, and, if most of the participants required appear available, book a meeting. With the above approach, the project manager minimizes the risk of having ineffective meetings by avoiding known scheduling conflicts.

> **Note**: It is important that all meeting facilitators complement this approach with adequate meeting facilitation training techniques to make each meeting successful.

5.2.2 Recurring Program/Project Update Communication Artifacts

One approach to effective communication is standardizing the way you communicate. The communication plan should include a recurring meeting for various types of status updates. These updates are necessary to discuss progress, identify and manage risks and issues, and determine the way forward while considering constraints and competing priorities. On larger programs, the sponsor is less likely to attend all detailed scrum or scrum-of-scrum sessions. Instead, separate meetings may be scheduled with the program and customer executive teams to discuss both status and strategic initiatives. Some programs prepare presentation decks to structure discussions and facilitate meetings with the core sponsor, while others use collaborative tools. Regardless of the tool used, a few standard items should be covered in these meetings:

a. **Program Health and Update Summary**

On smaller programs, compiling this information may be relatively straightforward. However, as team size grows, it becomes essential to determine the most critical items that need to be communicated to the sponsor and customer management team. A presentation slide or automated web-based tool may include a stoplight-type indicator to quickly assess the health of the program. Additionally, a summary of the primary activities completed since the last meeting and the upcoming tasks should be provided.

b. **Action Items**

A list of action items required by any of the participants. These are usually at the executive level. Are there any critical decisions that need to be made by the customer? Have all the items discussed during the last session been completed?

c. **New Requests**

One of the most challenging tasks of a program manager is managing scope creep. Clarifying expectations and trade-offs, as necessary, for any new requests should be included as an agenda item. This allows the program to gain a better appreciation of the value of strategic initiatives, confirm priorities, and discuss impact to help better manage scope. A program manager who does this well will then have a track record of completing projects on time and under budget.

d. **Additional Program Highlights**

On larger programs, deputy program managers or key team leaders may also be present to provide more information into parts of the program. It is best to summarize each of these to one page and have a standard approach to capture and communicate the status, issues and risks with mitigation for each item, and the key milestones and external dependencies. The more a program office can automate this information using collaboration tools, the more efficiency can be driven by removing this overhead burden on the team to focus on project implementation and delivery.

5.2.3 Requirement Capture

The success of every development effort begins with clear requirements. Customers of a new system usually have a general idea of what is expected, and these expectations evolve throughout the project lifecycle. Requirement management can be a full-time job, depending on the size and complexity of the program.

A business analyst's (BA) role on a technical team is primarily to work with the business sponsor and customer to clearly define the scope, specific requirements, business value, and priority of each capability. The BA is responsible for translating business needs into a format that the development team can structure into their development activities while also ensuring that clarifying questions are addressed.

Over the years, various techniques have been used to capture requirements, including the use of 'Shall Statements' (e.g., 'The system shall...') and, more commonly, User Stories. User stories are a concept used in Agile development to capture requirements effectively.

The typical flow for developing new requirements includes identifying an approved Change Request (CR). The change request begins with the identification of a solution, such as an application, dashboard, or report. The solution is then broken down into features and further decomposed into capabilities. These components can be used to create an Epic, which is entered into a ticketing system. The Epic is further divided into user stories that a developer can complete within a sprint cycle. These stories can be broken down into tasks and subtasks as needed.

The key point to remember is that user stories should be small enough to be completed within a single sprint cycle. Using this approach enables the capture of additional metrics regarding team performance. For more information on Agile, see **Sections 6.2 Agile: The Product Backlog and 8.2 Data-Related Projects - Shift to Agile.**

5.2.4 Requirement Confirmation

Before new capabilities are released to the customer, the acceptance criteria for each capability should be confirmed by the Product Owner (PO) or product sponsor. Communication is particularly important during this phase to make the most of the feedback by future customers of the system. Some organizations have formal user acceptance tests. Other techniques that can be used to familiarize the user base with new capability include user training and having an environment that can be used to play with upcoming capability. Either way, it is important to have a process in place for efficiently capturing feedback for requirements. The two important questions for discussion related to this are: what is the best way to communicate with our customer, and how best do we capture the feedback provided by our customer.

5.2.4.1 Communication to Customers

The project manager should work directly with the Product Owner, Business Analyst, and Test Lead when performing the new capability validation to ensure the process is clear. This communication should address questions the customers tasked with this activity will likely ask including:

1. When does the initiative start?
2. What is being validated?
3. Where am I conducting my validation?
4. Are there multiple system roles and, if so, which roles will I require for validation?
5. How long do I have to complete my validation?
6. How am I to log or document observations or findings?
7. Will we have an opportunity for periodic review of our findings?

5.2.4.2 Capture of Findings

Technical teams are caught up using terms like "issue" or "defect" when discussing observations made by a customer while testing a new system or a new capability of an existing system. This is often driven by the customer making statements such as "I found an issue" or by technology where the ticketing system implemented refers to each item as an "issue". The problem with the 'issue' label is that the label can quickly lead to a misunderstanding regarding product quality from the customer and result in frustration from the technical team. Strong technical resources who thrive on building solutions often go beyond to satisfy the customer. As a project manager on these types of projects, it is important to have a consistent definition and set expectations at the onset of such testing or validation initiatives.

> **Anecdote**: Instead of using terms like "issue" or "defect" for changes proposed by the user community testing the application, it is best to use terms like "finding" or "observation" captured in a separate independent log.

This approach has several benefits. First, it allows the user community involved with testing to capture all types of changes including understanding the rationale for the change, business value, and priority. The business further communicates if a finding is a change that is necessary before the product can be used, a change request that has a temporary workaround, or a change request that is minor/cosmetic, and be implemented when the development team has an opportunity. Second, the project manager can work with the development team to better understand if the finding is a defect, code developed that does not meet the acceptance criteria, an unintended defect, a new change request based on clarification of a business requirement; or a new requirement that the customer would like to add.

Meeting with the development team and the user community testers helps to facilitate communication on these observations and enhance the solution while protecting the scope and integrity of the project. Triaging the findings to determine and to log specific defects and issues for resolution, and to log new change requests for future backlog grooming or requirement sessions is necessary. This is particularly helpful when there are major business process changes that occur to complement a new capability. Prioritizing the findings is also important as it helps with the implementation strategy of future changes based on the value to the business and customer community.

5.3 Scope Management

One of the most difficult responsibilities of a project manager is managing scope creep. The project manager is naturally in a trusted advisor role and, as such, must develop good relationships with the customer while supporting the team. Some contracts are written with very specific deliverables and timelines while others leave a bit more room for interpretation. The organizations who utilize a clear scope management approach can benefit if created correctly. From a customer perspective, the approach may allow for reduced acquisition costs and minimize the number of vendors and associated overhead required to complete relevant solutions that meet their organizational objectives.

From a contractor perspective, effectively managing scope allows opportunities to add value and proves to be a strategic partner by delivery of the most value-added services. Customers always want to maximize the value of solutions by minimizing costs while contractors try to provide value-added solutions while increasing their profitability for solutions. As a program or project manager, these two ideologies often conflict and the PM must be in a position to navigate and balance the value expected by the customer with the margins expected by the contracting company. Open and honest communication in a trusted advisor relationship is best for someone in this role. A simple value proposition template that includes all inflight initiatives minimizes schedule risk by having a clear understanding of all inflight initiatives, estimated costs, impact to other initiatives, and the value that each provides. Including program constraints, a clear prioritization of work activity can be communicated and coordinated, allowing for optimized resource allocation of the inflight initiatives given the program constraints. It is also important to make sure all work is visible including several types of work typical on technical projects: operations, maintenance, and new capability. One approach that many organizations are moving toward is the use of a DEVSECOPS or SECDEVOPS model, where development, security, and operation work together as one unit instead of separate teams.

"The purpose and intent of DevSecOps is to build on the mindset that everyone is responsible for security with the goal of safely distributing security decisions at speed and scale to those who hold the highest level of context without sacrificing the safety required" describes Shannon Lietz, co-author of the "DevSecOps Manifesto." (Lietz, 2016)

6 SOLUTION DEVELOPMENT

The objective of the Solution Development section is to create an environment where developers can quickly access the information they need to rapidly develop products.

Primary Goals:

- Importance of standards
- Developing with a security mindset
- System Development Life Cycle
- Using Agile techniques for solution optimization

6.1 Development Standards

A recurring theme in this reference publication is the need to utilize standards to build high-performing teams. None is more critical than setting the development standards for a large solution development effort. Standards in a solution development initiative provides an environment where developers can quickly access the information they need to reference while developing products. The goal is to understand the importance of standards through onboard training, developing with a security mindset, managing risks, using proven techniques for solution optimization, and measuring development efforts.

6.1.1 DevSec: Developing with a Security Mindset

Technology has evolved significantly over the past twenty to thirty years. The rise of using technology by bad actors has increased the need for evolving technology to keep ahead of security infrastructure and underlying code of modern solutions. Not so long ago, requirements were captured; code was developed and tested and then deployed into a standard system development life cycle. Security was a separate group who reviewed code after the testing phase and as incidents came up in production. The primary task in general commercial application was vulnerability scanning and remediation. With technology and the bad actors who use it becoming more sophisticated, there was a need to formalize a working committee and establish a certification due to the growing demand of cyber security professionals. The International Information Systems Security Certification Consortium (ISC2) was established as a non-profit in 1989 and launched its first certification, Certified Information Security

Systems Professional in 1994. "The ISC2 serves to educate, empower, embrace, and engage our members through every step of their careers." (About, n.d.) The organization reached 10,000 members in 2002 and, as of 2023, surpassed 500,000 members. Now security is an integral part of the development process starting with security coding standards. There is an industry that dives specifically into the topic of security in application development and the following section scratches the surface of items one should consider in the realm of security management.

6.1.1.1 Security Coding Standards

Maintaining a standard list of security coding standards is an essential part of a development team onboarding and important highlights should be reviewed with the entire team periodically. Over time some of the leading practices become apparent so it is best to understand and provide guidance to everyone on the team. Some of the leading practices to ensure that code is secure include:

- User Input Validation: Confirm that all input fields are validated to prevent injection attacks including Structured Query Language (SQL) injections, Cross Site Scripting (XSS), and command injections
- Authentication and Authorization: Confirm the proper authentication mechanisms are in place and the access controls are enforced to prevent unauthorized access
- Data Encryption: Confirm that all, or at least sensitive data is encrypted both at rest and while in transit
- Security Headers: For systems available on the web, secure headers such as the content security policy
- Code Practices: Implement least privilege, zero trust, principles, and confirm avoidance of both hardcoded credentials and parameterized queries

- Security Testing: Perform regular, periodic security testing such as static code analysis and penetration testing to identify and remediate vulnerabilities.
- Error Handling: Review error handling routines to make sure sensitive data is not inadvertently included in error messages exposing security vulnerabilities
- Monitoring and Logging: Confirm effective implementation of monitoring to detect, capture pertinent details, and respond to security incidents
- Secure Configuration: Confirm that the use of default configuration is minimized and all configurations are properly secured.
- Manage Dependencies: Perform regular updates of patches to remediate known vulnerabilities across all applications, databases, systems and networks.

Most of the updates to the standards are from the feedback loop of the vulnerability monitoring by the security team. There are numerous templates and checklists available by organizations such as the National Institute of Standards and Technology (NIST). NIST has a Computer Security Division that, according to their website, uses their Computer Security Resource Center (CSRC) 'to encourage broad sharing of information security tools and practices, to provide a resource for information security standards and guidelines, and to identify and link key security Web resources to support the industry.' (NIST Information Technology Laboratory - Computer Security Division, n.d.) Implementing a framework such as the NIST Cyber Security Framework, is used to understand, assess, prioritize, and communicate cybersecurity risks:

- Identify
- Protect
- Detect
- Respond
- Recover
- Govern

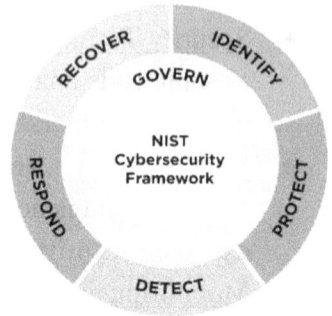

6.1.1.2 Knowledge/Training - ICS2

Security management in application development is critical in developing trusted solutions in the organization. Understanding the organizations who drive policy and standards; and certifies individuals in this area, is important to be successful in managing technical teams. A few of the area's to become more familiar with is found in the sections that follow.

Anecdote: The knowledge and training available for security management is extensive and can become a bit overwhelming. Fear not, remember we are life-long learners who are always looking for areas to grow. In my case, I once had to take on a small technical team that was responsible for important and sensitive data for a government organization. I was surprised to see that, although the resources were very knowledgeable and good at what they do, they were not credentialed. Knowing that the government was going to be enforcing credentials as part of their future contracts and that we only had a couple years before our recompete, I chose to lead by example and obtained my Certified Information Systems Security Professional (CISSP) certification from the International Information System Security Certification Consortium (ISC2). I also made obtaining certification part of each team member's performance goals for the year and focused on those who also required a CISSP level certification for their position. By obtaining my CISSP first, I quickly gained the respect and trust of the team, and I pushed them to align their goals (individual technical certification in their area of competency), to the goals of the organization. This approach allowed us to make a strong case for our recompete a couple years later. The additional motivational factor was the team understanding that, even if we did not win the contract, by continuing their professional development journey, they were each positioned to quickly find new and exciting opportunities to continue their life-long learning journey. This kept down the stress levels to a more manageable level and allowed the team to focus on the implementation and maintenance of this critical solution.

6.1.1.2.1 The National Institute of Standards and Technology

The National Institute of Standards and Technology is part of the United States Department of Commerce with a mission of "Working with industry and academia to enhance economic security and improve our quality of life." (National Institute of Standards and Technology, 2024). This institution has a wealth of resources available for both public and private sectors. Since the mid-1990s, the institution has provided computer, cyber, information security, and privacy to the public through its Computer Security Resource Center (CSRC) (NIST Computer Security Resource Center, 2024). On their website, you can review the latest cybersecurity framework and actively participate in reviewing and commenting on any of their draft publications. One of the themes driven into everyone in cybersecurity is that we each have an important role in securing our information assets. Anyone pursuing a technical career should know about this organization and review publications as an important source on keeping up with industry changes.

6.1.1.2.2 NIST: Zero Trust Architecture (ZTA)

A ZTA enterprise's cybersecurity plan utilizes zero trust concepts and encompasses component relationships, workflow planning, and access policies. Therefore, a zero-trust enterprise is the network infrastructure (physical and virtual), and operational policies that are in place for an enterprise as a product of a zero-trust architecture plan (Glossary, 2024).

6.1.1.2.3 NIST: Development Standards

NIST published the Interagency or Internal Report (IR) 8397 - Guidelines on Minimum Standards for Developer Verification of Software (NIST IR 8397 Guidelines on Minimum Standards of Developer Verification of Software, 2024). The document recommends techniques that are broadly applicable and form the minimum standard. It recommends the following techniques:

- Threat modeling to look for design-level security issues
- Automated testing for consistency and to minimize human effort
- Static code scanning to look for top bugs
- Heuristic tools to look for possible hardcoded secrets
- Use of built-in checks and protections
- 'Black box' test cases
- Code-based structural test cases
- Historical test cases
- Fuzzing
- Web app scanners, if applicable
- Address included code (libraries, packages, services)

Understanding the frameworks and guidelines put out by government agencies allows you to adjust your development standard to include and highlight specific dos and don'ts (no hardcoding passwords, changing default passwords, etc.), thus making your standards the best possible for your environment.

6.1.1.2.4 Department of Defense Security Technical Implementation Guides (STIGs)

STIGs are based on Department of Defense (DoD) policy and security controls. The implementation guide is geared to a specific product and version. The guide contains all requirements that have been flagged as applicable for the products which have been selected on a DoD baseline. (Glossary, 2024)

6.1.1.2.5 Plan of Action and Milestones (POAM)

A POAM document identifies tasks needing to be accomplished. It details resources required to accomplish the elements of the plan, any milestones for meeting the tasks, and scheduled milestone completion dates. (Glossary, 2024)

6.1.1.2.6 Zachman Framework

High performance teams utilize standards and frameworks to work more efficiently. Practitioners and educators spend careers trying to determine how to do things smarter, better, and faster. Understanding leading practice frameworks helps to optimize high performance teams. One such framework that is prevalent in technical solutions is the Zachman Framework™. This framework is a tool which can be used to classify an organizations architecture.

> "The Zachman Framework™ is a schema - the intersection between two historical classifications that have been in use for literally thousands of years. The first is the fundamentals of communication found in the primitive interrogatives: what, how, when, who, where, and why. The integration of answers to these questions enables the comprehensive, composite

description of complex ideas. The second is derived from reification, the transformation of an abstract idea into an instantiation that was initially postulated by ancient Greek philosophers and is labeled in the Zachman Framework™: Identification, Definition, Representation, Specification, Configuration and Instantiation." (https://zachman-feac.com/zachman/about-the-zachman-framework)

6.1.1.3 Scanning with Every Sprint/Release

Agile teams with a SecDevOps or DevSecOps mindset rely on tools that can scan as part of every sprint. Scanning small pieces of an overall solution has some limitations and provides a different opportunity for monitoring specific types of security vulnerabilities as part of the code set. Implemented during the sprint, scanning software allows the testers and developers to note trends and to learn more quickly about the correct way to build securely. The major advantage of this approach is the speed of learning which inherently affects future code development; in particular for a team that reuses standard code blocks for consistency. For example, if a new team member builds a code block that has a hardcoded password or parameter driven component that is missed during a code review, the security scanner will identify this for remediation. The code block can be secure and the vulnerability risk minimized. The team can also discuss and reemphasize the development of security standards as part of the retrospectives. As a result, within two to three weeks, the team is aware and alert of how to make the code base more secure.

Alternatively, if the security vulnerability review is only completed when the code is being released, it could take months before it is identified. The vulnerability may be applied in multiple areas before the issue is identified since the developers are unaware of the vulnerability. This builds unnecessary technical debt and takes time away from implementing new functionality.

The problem may not seem like a big one, however, many independent problems or one off's can quickly snowball into technical debt that has to be remediated. This may result in a difficult choice of delaying delivery of the solution or delivering less functionality.

The ideal solution is to integrate multiple security scans into the development pipeline. Understanding the tools available or architecting a security model that utilizes multiple tools, even if there is overlap, is important. This overlap allows the security team to have a more complete security management solution because each vendor may not use the same technique in scanning. Development pipeline security considerations should include both identification and remediation security techniques. Maintaining an automated security solution that combines interactive security testing with run-time application self-protection is a leading practice. Static Application Security Testing (SAST), Dynamic Application Security Testing (DAST), Interactive Application Security Testing (IAST), Secret Detection, Dependency Vulnerability, bug checks, code smells, duplicate code identification, and Common Vulnerabilities and Exposures (CVEs) techniques combined at various stages of the development release pipeline, minimize overall security risks. CVEs is a list of known vulnerabilities for code bases. There are national common vulnerability databases maintained by organizations such as NIST. Major application providers also have a list of security bulletins that are published as soon as they are aware of vulnerability risks in their products. The Security management teams should be aware of all bulletins with any application in the organization's suite of approved products. The open community organizations such as Open Worldwide Application Security Project (OWASP), a nonprofit foundation that works to improve the security of software, provides good information like the lists of the top vulnerability that can help educate a high performing security management team.

6.1.1.4 Release Expectations

Most solution development teams perform formal testing prior to releasing code. It is important to establish the rules of release and draw a strict line. For example, many government organizations will not allow code to be released with any critical or high vulnerability findings. This may seem obvious; however, when the pressure is on to deliver capability that meets an important mission, often management has to make tough decisions on the potential for breach vs the implementation of an application that may save lives in the field. Setting hard rules, however, helps to set the appropriate balance and forces the team to act with a security mindset. Understanding the significance of vulnerabilities in a production environment allows you to better appreciate the DevSecOps approach of solution development. Building technical debt because of unsecured code can put lives at risk if needed capability is delayed in some industries. With time and experience, a management team can use the most suitable processes to integrate security into their development model. Continuously educating the team on leading trends should become part of the strategic objective of the management team.

6.1.1.5 Monitoring and Remediating Vulnerabilities

There are bad actors who are constantly working to gain an advantage the easy way, by stealing intellectual capital via unsecured code; so on the operations side, the security team focus is on blocking attacks. This is accomplished through numerous techniques, including implementing

Web Application Firewalls (WAF), Intrusion Detection Systems (IDS), Intrusion Prevention Systems (IPS), and Runtime Application Self-Protection (RASP) solutions.

Anecdote: I recall working on a team with one of my strong system administrators who met with me at least once a week to talk about security and show evidence that bad actors were actively trying to gain access to some of our secured systems. Luckily, I knew we had a strong team who was passionate about our mission and would not let this occur. We worked together on a solution that strengthened our systems security posture and continued to actively monitor for the duration of the contract. This experience was the major reason I felt the need to obtain my CISSP and I encouraged the team to become more aware and prepared from a security perspective, by making security a priority and establishing it as part of both the team and individual performance goals. The team became smarter in each of our areas. As the DevSecOps concept was becoming more prevalent in the workspace, monitoring and remediating vulnerabilities now takes a full team of resources on some programs.

The Information Systems Security Officer is the role that is ultimately responsible for creating and enforcing processes and procedures performed by both the development team, as they develop new code and release to production; and the administrative team who is responsible to make sure applications are patched or upgraded to maintain an acceptable security posture. Given the correct tools to monitor via dashboards can be an effective way to efficiently remediate across teams of various sizes. The larger the team, the more important it is to optimize supporting mechanisms like security dashboard solutions, that align with the goals of the security organization. A schedule of recurring security activities should be created and maintained by the operations support team. This schedule may include items such as the upgrade schedule for your major vendors, patching activity, major application maintenance, and any other recurring dates that may be beneficial for quick access by the operations support team. This recurring schedule becomes a good training tool for onboarding new resources on these teams.

6.2 Agile: The Product Backlog

Having a defined list of development work items required to complete a solution is necessary in any system development life cycle. The clearer the requirements, the more likely the solution will be developed on schedule and on budget. Organizations who implement an agile development life cycle realize this as one of the most important benefits of agile. The Product Backlog usually refers to the list of new features, bug fixes, infrastructure changes, or other activities that a team may deliver to achieve a specific outcome. Items in the backlog include epics, capabilities/enablers, features, and user stories (Agile Glossary, 2024). The Scaled Agile Framework (SAFe) is a good resource for building competencies in this area. SAFe 6.0 is a knowledge base of proven, integrated principles, practices, and competencies for achieving business agility using Lean, Agile, and DevOps. The competencies covered in SAFe are included below (About, 2024):

- Lean-Agile Leadership
- Team and Technical Agility
- Agile Product Delivery
- Enterprise Solution Delivery

- Lean Portfolio Management
- Organizational Agility
- Continuous Learning Culture

One differentiator of SAFe is that it breaks the product backlog into distinct sets of backlogs based on granularity. The portfolio backlog includes the epics: the solution backlog includes the capabilities and enablers; the program backlog includes the features; and the Team backlog contains the most granular user stories and other items that a team works on.

6.2.1 Grooming and Pruning

A maturing agile program will ultimately lead to a growing backlog as requirements capture outpace the ability to develop capabilities or changes in the system. As a result, there is a need to periodically review the backlog for grooming and pruning of backlog items. This periodic review provides the opportunity to ensure a high-quality backlog by checking amongst other things:

- All tickets are broken down to the correct level of specificity for the stage in the backlog

The tickets in the backlog go through several levels of refinement as the work moves from Epics to Tasks and Sub-tasks. Tasks are also broken down into technical and functional tasks. Providing the team with a standard way of documenting the tickets from original creation to the final sprint refinement allows the team to operate most efficiently. This may start with a high-level understanding of the capability on the Epic level which is then broken down by the team into tasks that can be measured. Before user stories and its tasks are added to a Sprint, the leading practice is that the tasks are broken down to a level that meets a few criteria:

- Try to keep each sprint to one to three weeks
- Ensure all tickets are groomed before each sprint begins
- Avoid large user stories that cannot be completed in one sprint
- Tasks should be sized to be completed by an individual within a day - so two to eight hours is the ideal task sizing for a Sprint. This allows for demonstrating actionable delivery on a daily basis and helps the scrum master in reassignments.

Team building and consensus exercises can be used to engage the team in estimating the size of tasks. This can be a fun learning experience if you use blind surveys as part of the technique. This technique requires all team members to actively participate in the sizing of tasks. Based on the results, a brief discussion on the outliers allows engagement within the team to understand different perspectives and may even allow the scrum master to make better assignments based on outlier explanations. At the same time, it allows each individual scoring to better understand the rationale of other teammates, which may improve their logic on determining individual task sizing. What results is the continuous improvement of the team in estimating technique.

6.2.2 Sprint Planning

Now that your backlog is pruned and groomed, it is time for sprint planning. To make the best use of the product backlog, it is good to understand both the functional priority and technical complexity of the work in the backlog. Using these two metrics, along with historical measurements like a team's velocity, you can create a good sprint plan.

1. Review the sprint calendar and make sure that you have accounted for **holidays** and any **personal time off.**
2. Consider how much time will be spent focused on **non-development activity**. Activities such as training, meetings, etc. are usually included in this allocation. A general guideline is to reserve **15%** for other administrative type activities. This allocation can be reassessed and modified as the processes mature.
3. Based on Product Backlog Stories, **create or update the subtasks** of each including the estimated time to complete.
4. Confirm the **allocation** of each resource and check for over and underutilization.
5. Review the resource allocation to ensure the estimated time is less than the allocation.
6. For mature teams, **review** the teams and individual average **velocity** to highlight any trends and anomalies.
7. **Review** all the stories for each team member **with the team**. If each of the individual technical members works with the business analyst ahead of time, this can be done in typical scrum fashion or as a quick post-scrum activity without the product owner. Confirm that all stories have associated subtasks and that all subtasks have estimates.
8. Finalize the team's **commitment** and review the level of commitment as a team to confirm that it is achievable even with typical add-ons or 'drive-bys' that may occur on some teams.
9. **Provide** the team commitment for the sprint to the **product owner** for approval.
10. **Start** the sprint.

It is very important that each team member complete the subtasks associated with each user story including the estimated hours for each subtask as part of this planning activity. The estimated hours and actual hours are very important for future planning and for determining high quality key performance indicators. Actual hours should be completed daily and aligned with timekeeping for measurement that is more accurate. Some team members may delay entering the hours until the task is complete or just before the Sprint closes. This can lead to less accurate measurement and should be avoided. At the onset of the development effort, one tip is to include the rule of daily subtask entry as part of the initial instructions and team agreement.

6.2.3 Sprint Execution

Effective sprint planning is the key to good execution. Several heuristics should be implemented as part of the execution. Once a sprint has started, there should be no or very limited items added to the sprint.

Anecdote: If the program has an issue with frequent disruptions to the sprint that cannot be avoided, tasks should be allocated for drive by requests to plan for unexpected events. Start by allocating 15% of the sprint capacity to these drive by requests and modify depending on need. In some cases, this is not necessarily a bad thing as it may help support the benefits of Agile, however, the Scrum Master should ensure it is not affecting the team dynamics. The length of the Sprint should be reconsidered if this is problematic. If you have a three-week sprint cycle, you may want to shorten to two or even one week. It is much more difficult for a client to say, I can't wait one week vs I can't wait three weeks for a new capability.

In agile, leading practice is that nothing is added to the sprint. If a sprint is short enough, every two weeks for example, any new request will go into the backlog, which is then reprioritized for use in planning the next sprint. In reality, particularly in a DevSecOps model, it is not quite that simple. In a case where the development team also serves as a Tier three or Tier two-support model, customers may have urgent fix requests that the team needs to troubleshoot and fix. Over time, these 'drive-bys' can be measured and incorporated into your sprint planning. If the team spends thirty percent of their time on customer support activities, these should be captured in a separate story with the estimated allocated time. Although this approach helps, the nature of this type of model provides variability, as some customers may request more support in one sprint over the other. Actively tracking this activity is the best way for better estimating. Using tools like burn down charts and other sprint health mechanisms should allow the team to see how they are tracking according to plan. It is best to have the tracking tool clearly visible so the team can auto correct during the sprint as needed.

It is easier for a team to surge in one area early to help, than to wait until the last couple of days of the sprint to realize they are not going to make their delivery commitment.

Anecdote: The scrum master's primary role is to provide leading practice, ensure compliance to agile processes, and remove obstacles from the team. A typical desktop setup includes three different consoles. The first console controls the meetings that the scrum master facilitates. The second and primary screen should never be used for sensitive information and the background of the room should always be in a ready state in case you need to go on camera. The third screen is the Communication Central screen, and it is used to facilitate continuous communication with the team. Rapid note keeping and continuous improvement in facilitation skills helps the scrum master successfully manage high performing teams.

Web meeting Central
- Meeting Controls – Mute/Video
- Meeting Chat

Primary Presentation Screen
- Camera centered
- Never used for sensitive items
- Background always ready to go

Communication Central Screen
- Email
- General Chat
- Notes

6.2.4 Sprint Retrospective

Retrospectives are one of those ceremonies that are often not appreciated by a technical team. The technical team wants to build and create solutions. Retrospectives focus more on soft skills and understanding the importance of process improvements to make the team function better collectively. In a typical retrospective, the scrum master goes around to each team member asking four questions:

- What went well?
- What could have been better?
- What should we continue doing?
- What should we do differently?

Along with these questions, specific action items are captured and categorized on the project or program level. The problem with this textbook approach is that it is hard for some individuals to provide feedback for numerous reasons. Think about the last time you had a whiteboard session. Usually, the most difficult part is filling the whiteboard. It is much easier to look at something and provide feedback on what is correct, incorrect, or can be improved. Similarly, the retrospective can be seen as a 'checkbox' ceremony where everyone says how great the team is doing but have little to no comments on improvements or only comment on those most critical things that the team remembers. This is no fault of the team. High performing teams go through a lot during a sprint cycle so it is often difficult to recollect outside of the last fire drill or major development activity.

Anecdote: Capturing retrospective review items during the sprint is the single most effective way of changing the mindset and making this agile ceremony most effective. Like the white board session, if provided with specific retrospective items to review and update, versus having the team come into the meeting with general open-ended questions, leads to better discussion. The added benefit of this approach is that once team members start reflecting on these areas, they open up with other ideas and unknown items to the broader team which allows more effective capture and provide good action items for improvement.

7 RESOURCE MANAGEMENT

The objective of the Resource Management section is to focus on managing the staff and implementing methods for effective use of risk and quality management to better support customers.

Primary Goals:

- Staff Management
- Governance
- Risk and Quality Management
- Implementation of Customer Support
- Measuring Team Effectiveness

7.1 Staff Management

Good program management does not happen by accident. It starts with creating an infrastructure and team structure that maximizes productivity. Setting up a framework for success can begin by exploring three areas: the program management office, the management team, and the team structure.

7.1.1 Program/Project Management Office

Before we dive into the program management and office discussion, it is important to clarify the definition of a project and a program.

> According to the PMBOK® Guide —Seventh edition (PMI, 2021, p. 4) the definition of a project is 'a temporary endeavor undertaken to create a unique project service or result.' Projects are temporary and close down once the work chartered is completed.

> The definition of a program as given in The Standard for Program Management —Second edition (PMI, 2021, p. 4) is 'Related projects, subsidiary programs, and program activities that are managed in a coordinated manner to obtain benefits not available from managing them individually.'

Based on these definitions, many organizations have a portfolio of projects that they regularly undertake, yet not all companies have a project management office. According to some reports, up to 84% of companies have at least one project management office. Important differentiators as to what make project offices successful are well documented from experienced subject matter experts

"One of the top differentiators of success is how well a PMO is embedded within an organization" (PMI, 2012), Dr. Aubry says. She lists four factors that determine this level of integration:

1. **Collaboration**: The PMO should encourage collaboration between project professionals and functional departments.
2. **Recognition of expertise**: Do the project professionals working with the PMO improve the level of respect the project management receives within the organization? This should also influence who works in the PMO.
3. **The mission is well understood**: Do those outside the PMO know its purpose?
4. **Support from upper management**: Is there an executive champion who will not only communicate the mission, but will work to gain engagement from stakeholders?

Understanding the success criteria up front allows the program manager to optimize the necessary artifacts that will be required to communicate for success. The project management office's primary goal is to support the team to achieve success.

PMI does a really good job keeping up with the changes in the project management profession. Every year, they publish their "Pulse of the Profession Report". Over the past few years the organization has been shifting the focus from project management success alone to include project success. Reading this type of publication allows you to understand trends and changes in the project management industry. These trends and changes are good discussion topics to take back to the organization teams for discussion. Understanding industry trends allows the team to adopt by looking forward and tweaking their processes as part of continuous improvement efforts. In 2024, PMI's

Pulse of Profession Report focus was on Maximizing Project Successes through Business Acumen. This completed their exploration of the PMI Talent Triangle™ following Ways of Working in 2024 and Power Skills in 2023. The key takeaway on this year's report was demonstrating how business acumen leads to better project outcomes.

Individuals managing high performing teams keep up with industry trends and look for opportunities to incorporate lessons learned from research as part of their continuous improvement processes.

7.1.2 Management Team

High-performing teams require leadership that knows how to get the job done. It is exceedingly difficult to optimize a high-performance team successfully if they do not have the leader who is setting the direction, providing the required support to maximize the value of each team member, and protecting the team as needed. Trust and competence are two attributes you will find in any leader of such teams. In his 2000 Harvard Business Review article, Leadership That Gets Results (Goleman, 2000), Daniel Goldman first introduced a collection of six leadership styles that leaders should have in their arsenal. Using these styles effectively gives a leader flexibility; however, it is hard to implement. The good news is that these styles can be learned, and with practice, can also be used to maximize the value of the team's performance. "A leader's success depends on the productivity and effectiveness of the people who work for them," Goldman says. "You're shooting yourself in the foot if you use a style of leadership that's counterproductive to their performance." These leadership styles are found below:

- **Coercive leadership style**, which entails demanding immediate compliance.

- **Authoritative leadership style**, which mobilizes people toward a vision.
- **Pacesetting leadership style**, which expects self-direction and excellence.
- **Affiliative leadership style**, which centers on building emotional bonds.
- **Democratic leadership style**, which involves creating consensus.
- **Coaching leadership style**, which focuses on developing people for the future.

Today, this collection of leadership styles remains relevant. In her article, 6 Common Leadership Styles – and How to Decide Which to Use When (Knight, 2024), Rebecca Knight states, "Even if you're naturally introverted or you tend to be driven by data and analysis rather than emotion, you can still learn how to adapt different leadership styles to organize, motivate, and direct your team." Although in this publication, more of this subject will not be covered directly, if your interest is becoming a leader in managing high performance teams, understanding the different leadership styles and learning how to implement them effectively will be an important part of your professional journey.

7.1.3 Team Structure

Structuring the team for success is another role of the project manager. One approach to start the assessment in this area is to understand how effective teams are structured. In general, five and seven resources is the ideal team size. Once you get beyond seven resources, there are diminishing returns on team performance and effective management, which can be disruptive to a team. It is important that each team member be given the care required to perform optimally. This requires a concerted effort to make sure everyone has the support they need in growing professionally, communicating concerns, and operating in a safe and professional environment where everyone is working together on a specific goal or mission. On some larger programs, the project managers may have over a dozen direct reports. This is only acceptable if it is on a temporary basis to help during transitioning. New resources may be on boarded with an initial manager until they are fully on boarded at which time, they change managers. This may also occur with changes in the management team or in a surge of hiring/recruiting. Over the long term, it is important to try to set a goal of no one specific person having more than five to seven resources for more than three months. By realigning the team structure to have the optimal five to seven resources during the life of the program, the management team optimizes the chances of continued high team performance. There are also additional techniques that should be considered regarding the organizational structure and management style of the teams.

Anecdote: On one program, I was skeptical when we decided to try a new management structure where none of the technical team leads would have direct reports, and instead, we would bring in project managers who would take on the Human Resource management of the team members. Frequently, technical managers often take on management roles without sufficient training. As a result, they see the human resource management role as less of a priority compared to delivering the work. Once the new management structure was implemented and the technical leads understood they still had the authority and ownership of technical delivery and guidance, they warmed up to letting go of having 'manager' responsibilities for their team. The Project Manager worked collaboratively with the Technical Lead and Business Analyst Lead to create a cohesive management structure that worked for over five years until the program was completed. Communication played a key role in the success of the program. First, the role of the project manager, technical lead and business analyst were clearly defined. After gaining consensus, the roles were communicated to the staff. This answered the question "From whom do I take direction?" Second, the onboarding training document was updated with a few slides to address the new framework and showed how the three roles complimented each other. After seeing the new framework, the team quickly adjusted and saw the benefits of the approach. The technical leads were also happy they did not have to focus on timecards, regular formal touch points that had to be documented, and other administrative 'stuff'. The staff now had multiple individuals they could talk to allowing for their voice to be heard more often.

A similar approach is used on Agile programs where the trifecta of the scrum master, product architect, and business analyst helps to drive the work done on sprints. *See 8.2 Data Related Projects - Shift to Agile* for more information.

7.1.4 Optimizing a High Performing Team

Optimizing a high-performance team must be intentional from the beginning of a project. Every team member should have two things to enhance their success: an onboarding training plan and a mentor. The onboarding training plan should include all mandated training for the company, an overview of the program and how their role fits with the deliverables, and a plan for increasing complex activity for the first month of engagement. A mentor should also be assigned as the primary point of contact for any questions and to help with team norms or orientation.

- Every member of the team should be required to have onboarding prior to starting, that includes a good overview of the program/project mission and justification.
- Periodic reviews should set SMART goals, which include self-development goals that align with corporate and program objectives.
- Challenge those with the least knowledge by assigning tasks that are low priority and high value. If the application being developed has reference value fields, the new team member may be able to conduct research to draft the values and definitions. If there are new business processes that are drafted, have the new individual review it to make sure it makes sense to someone with limited familiarity on the topic.

> **Anecdote**: I use this exercise to improve the onboarding process by asking the person to provide feedback and proactively communicate improvements as they are onboarded. This allows the individual to better engage, not taking classes because they are mandatory, but also updating to answer additional questions that may not be apparent.

This approach offers several benefits:

- o The individual is proactively engaged
- o The onboarding process is continuously improving with every added team member
- o The more senior team members provide guidance and team building starts
- o Senior team members can be coached and continue to perform their own growth to higher levels in the organization
- o Typically, this approach saves time by having a resource who is engaged and more productive while freeing up senior resources from hand holding throughout the lifecycle of the onboarding process. Providing an understanding of roles, the individual teaches whom to go to for the right type of information more quickly instead of solely relying on their direct supervisor. Metrics can be predefined to see measurable changes via the increase in velocity or decrease in cost per story point.
- Organization Training: Companies need to be careful about the types of training required as part of the onboarding. Is specific training important for the role or the organization? If so, is it optional or required? If it is required, it indicates it is part of the

core value or training needed by everyone in the organization, a common understanding to level set the core competency or ideology the company stands for.

> **Anecdote**: I use this as a simple example of my personal experience where organizations would speak to the importance of core company values yet make training optional. Those who attend optional training programs are typically those who already understand the value of the training. I have also observed organizations that truly seeks to build leadership having the C-level suite provide the training. Larger consulting organizations typically hire subject matter experts from a diverse background (experience and education), which allows the organization to quickly disseminate consistent high-quality deliverables.

The challenge for the program office is to balance the training required to make good professional resources that complement the core knowledge of the broad set of resources, with the specific industry or technology technical training to improve the solution development across the team. The challenge is directly proportional to the program size, and setting the goals and objectives of training early in the program is important to be most effective.

7.1.5 Feedback Loop

A feedback loop is important to maintain the continuous improvement of the individual, team, or organization. It is difficult to distribute your time most efficiently if you do not monitor and understand the most effective ways to enhance the individual and team to align to the program, division, and organizational goals.

7.1.5.1 Personal Reflection

Personal reflection is important to continue professional growth. Everyone, life-long learners in particular, have continuously changing goals and objectives they want to achieve. From a career perspective, it may be getting that next promotion. From a development perspective, it may be learning a new technology. From a more personal perspective, it may be finding a mate, expanding the family, travelling, or starting a hobby. Each individual is unique and has their own set of values of what is important, which may change in priority over time. Given this fact, how do we optimize a team's performance? The key is to try to align personal goals with professional, team, and organizational goals. This brings back the importance of team building and having a strong management team whom the individual trusts. Many professionals try to keep their professional and personal lives separate. However, the more you know about individual values and interests, the more you can align each team member's personal and professional goals. The team member who has a personal interest in learning Agile and being certified, may be assigned to facilitate more of the formal ceremonies. The team member who has a personal interest in photography may be assigned to take photos at a team-building event. Another team member may be rewarded and be more appreciative of theatre tickets instead of a monetary bonus. Alignment of personal and career objectives allows you to not just know the database administrator who works on the team but also get to a level that really understands John Doe, an individual who happens to also be a database administrator. Understanding the team to this level helps the individual manager to manage better and sets up for the optimization of the team.

Anecdote: Every year during the holiday period between December 15 and January 2nd, I use the opportunity to take some down time, recharge and spend quality time with family and friends. I also reflect on my accomplishments and ways I could improve my personal and business relationships. Did I achieve what I set out to for the year? What should I have done differently? What can I do next year that is specific, measurable, achievable, relevant and timebound? Self-reflection allows me to redefine my goals for the next year and continue to be a better person.

7.1.5.2 Team Reflection

Over the years, the large consulting organizations have evolved the way they conduct their formal periodical performance reviews. Some perform quarterly reviews, and others annual reviews. In every case, the review is only as effective as the manager who is providing that review.

On a tactical level, organizations that implement an Agile development cycle use the Sprint Retrospective to better reflect on the work the team is doing. The retrospective answers four basic questions:

1. What went well?
2. What could have been better?
3. What should we continue doing?
4. What should we do differently?

Performed at the close of every sprint cycle, typically two to four weeks in duration, the team has a chance to identify and suggest improvements regarding tactical items while it remains top of mind. The retrospective also allows the team to reflect on how the changes captured in previous retrospectives helped to make the individual and the team better. It often takes a strong facilitator to make these meetings most effective. If not done correctly, individual team members may not find value in the exercise and it becomes a 'checkbox' meeting, one that they just attend because it is part of a formal agile ceremony forced by the program.

Seasoned teams who have worked together for many years are the biggest challenge with this mindset. They are used to doing things well together for a long time and they are so busy they do not take the time to step back and see the value of the ceremony. This is one of the most difficult areas for seasoned teams who are learning to adapt to agile. The individuals already understand the nuances of their team members and what it takes to make the team effective, and they are quick to praise and avoid providing negative feedback that may hurt the other team members' feelings. There are tools and techniques that a scrum master learns to overcome this mindset by showing the value of the team products, highlighting the wins, identifying some of the challenges that were overcome, how the team overcame them, and how to avoid such challenges in the future.

Capturing what went well and what can be improved can be captured as it happens on a daily basis and then consolidating for review by the team in a daily scrum at the close of the sprint is most effective. This approach provides an independent perspective of suggestions in each area or role and provides the individuals in the area an opportunity to agree or modify the suggestion. This also opens everyone up to start thinking about things to correct perception, or to add additional insights that may not be known by everyone on the team. It is easier for individuals to react to something tangible than ask the generic questions of what went well, what didn't, etc. This approach, although an effective and essential part of the ceremony, usually only captures the big wins and improvements. Having an agile mindset, however, allows you to better understand that even making multiple minor, often forgotten improvements can lead to big change over time. The organizations can have all the formal processes and procedures in place, however, if they do not also hold their management team accountable for providing feedback, it becomes less effective.

7.1.5.3 Organization Assessment

Those who fail to plan, plan to fail. This is one of the core tenets any project manager who is serious about their craft lives by. Organizations should start any initiative by performing an assessment as to where they are, where they would like to be at the end of the initiative, and to be true about what it will take to get there. They may include using tools and techniques like the SWOT analysis when considering moving into a new business or product line. Based on this information, they may conduct an assessment to better understand where they are, where they are perceived to be, and build a roadmap to guide them on their journey. Survey based organizational assessments are best done in conjunction with live interviews with stakeholders across the organizations. While the assessments can capture very good information that can be easily analyzed to determine a roadmap, nothing is more valuable than listening and observing stakeholder verbal and non-verbal cues. These cues help the seasoned professional with an opportunity to pull and explore threads that may not be readily picked up in an assessment where participants want to provide feedback and revise their stream of consciousness, before actually submitting the survey.

7.2 Governance (Technology, Data)

Governance is an administrative process often overlooked by strong engineering or technical teams whose focus is trying to get the work done. If a customer does not have a formal data governance process in place, one needs to be seeded. Many publications focus on building governance organizations for technical implementations. My first book, A Practical Guide to Implementing an Enterprise Information Management Program (Boschulte, 2010), provides additional details on this topic. The high-level process for building a good governance structure is included below:

- Identify those in the organization who are best to sponsor the team
- Identify those in the organization who have a direct vested interest in the success of the team
- Create a biweekly meeting for cross technical team communication
- Formalize the agenda with specific approval or agenda item for each of the participating teams
  - These groups will typically be director or mid-level resources who the sponsors lean on to answer questions. This same group will evolve into voting members of the formal governance board.
- Use the project to 'prime the pump' on communicating the value of the group. This requires a strong facilitator to manage the biweekly agenda so these important participants clearly understand their role, the importance of the group, and can 'easily' participate while providing good information across the organization.
- Provide examples of rogue groups getting ahead of policy and standards

- If formal business workflows do not exist, this is a good area to start.
  - Are the existing processes documented? Are those documents current and up to date? Has the subject matter expert for each division reviewed them?
    - **Note**: Keep in mind that customers are used to doing their jobs the same way and often overlook inherent assumptions or steps that are just a part of doing business. A senior facilitator or business analyst should be able to navigate leadership and knowledgeable experts and be a very good communicator to drive out nuances that are inherent in any evolving business workflow.
  - Based on feedback from the divisions and the solution being implemented, are there changes or new flows that should be reflected in a 'to be' process flow? If so, these should be included in the communication and training plan.
  - Throughout the life of the project, these business workflows should be constantly reviewed and refined with detailed step-by-steps that include additional details like timeliness, detailed steps or other important notes that may not be relevant to see on a high-level flow.

7.3 Risk and Quality Management

Risk and Quality Management is one area that can help directly affect the success of a program. The larger the program, the more robust a solution a program office requires to track and monitor a program's progress ensuring the team continues to minimize risk and implements high quality solutions. Good resources recommended to support this area are passionate enough about this competency to become certified. The Project Management Institute offers the PMI Risk Management Professional (PMI-RMP) certification, which highlights your ability to identify and assess project risks, mitigate threats, and capitalize on opportunities. In this capacity, you enhance and protect the needs of your organization (PMI, 2024). There are even certifications focused on risk in specific industries. CPHRM certified professionals offer expertise in minimizing risk in a health care setting through proven knowledge of clinical and patient safety programs, risk financing, and compliance with federal regulations (AHA, 2024). There are complimentary standards/frameworks such as Failure Modes and Effects Analysis (FMEA), a risk assessment tool that evaluates the severity, occurrence, and detection of risks to prioritize which ones are the most urgent. Likewise, Quality Management is a competency that can take a lifetime to learn. Those who are passionate in this area usually become certified in tools like Six Sigma and study the International Standards-Quality Management. ISO 9001:2015 specifies requirements for a quality management system within an organization. (ISO, 2024) Industry specific frameworks such as advanced product quality planning (APQP), is a framework of procedures and techniques used to develop products in various industries, particularly in the automotive industry. It differs from Design for Six Sigma in that the goal of DFSS is to reduce variation. (Lean Six Sigma, 2024) The important point here is that there are references and tools available for each person to become better in what they do. As an individual team member, it is easy to identify those who are driven to learn about their passion. Those individuals make a high-performing team and typically have a strong work ethic when supported by management. The leadership team needs to have the ability to recognize strong team members to help them continue to optimize their teams to achieve great things.

Career Note: The risk manager and quality manager roles are often combined on smaller programs. These are two different competencies and each can be a full-time role on larger programs.

7.3.1 Risk Management

Programs should always capture risks using a risk management tool. Many mature programs include a Risk Matrix, a five by five table that indicates the probability and impact of a risk. In the risk matrix, the cells of the table are numbered one through five for both the X-axis, probability, and the Y-axis, impact. Probability and index are well defined so that the team can select the best ranking based on these two criteria. A Risk Score is then created by multiplying the probability index and impact index. A well-defined risk management process will set a threshold to raise issues with the management team, as well as set a minimum number of fields and default risk values for quick entry.

Note: A program that has matured to this level provides additional insight including the monitoring of how many on the team are using default values and how those default values materialize into actual probabilities to assess changing the default value.

Risk	5	5	10	15	20	25
Impact	4	4	8	12	16	20
	3	3	6	9	12	15
	2	2	4	6	8	10
	1	1	2	3	4	5
		1	2	3	4	5
		Risk Probability				

Table 4: Determining Risk Impact

Example: A program sets a default value of risk with an impact of 3 and a probability of 3. Any risk that exceeds 15 is presented to the management team for awareness. Any risks that have a score higher than 6 may be reassessed on a periodic basis by the management team. As risks are entered, the team member entering the risk will place a likelihood (probability), and consequence (impact), on each risk. Multiplying the risk impact and probability gives you a risk score.

A similar table can be used to identify the risk by each risk category. For larger programs, this is a very insightful table that allows better risk management.

Risk Impact					
5	1	2	0	0	1
4	0	0	1	0	0
3	0	2	6	1	1
2	0	0	1	0	0
1	1	0	2	0	0
	1	2	3	4	5
	Risk Probability				

Table 5 Risk Summary Table

Based on the example provided in Table 5, there are two risks that will be reviewed by the executive team, ten risks that the program management team will closely monitor and there are nineteen risks identified in the risk register.

The Risk Matrix allows the management team to quickly assess the health of the program and how well the management team is doing on remediating. More mature programs that use automated tools for risk can also quickly report on the top five risks, the next five risks pending based on the impact dates, the and the latest risks to be entered.

Separate sessions may be held by the sponsor to complete a detailed analysis if the risk register grows too large. Regardless of the size of the register, having the risk matrix with the summary of these key metrics defined above will provide sufficient information for an executive touch point session.

7.3.2 Quality Management

Quality Management Systems (QMS) is one competency in which any highly performing team manager should be well versed. The International Standards Organization (ISO) is a good place to start. Their Technical Committee (TC) 176 (ISO/TC 176, 2024), has been a leader in the development of standards in the field of quality management systems and tools since 1979. There are seven core tenants of good quality management systems.

- **Customer focus**: At the heart of quality management is the core aim of meeting customer requirements and striving to exceed the public's expectations.
- **Leadership**: Successful leaders establish a unified sense of purpose and direction. They continuously create optimal conditions in which employees flourish, gaining motivation and professional satisfaction in the pursuit of quality objectives.
- **Engagement**: Inspiring and empowering people at all levels is essential to the value delivery process.
- **Process approach**: A well-structured, coherent system is an effective container in which consistent and predictable results can be achieved effectively and efficiently.
- **Improvement**: A hallmark of successful organizations is a continuous dedication to improvement, from product or service

quality to meeting or exceeding customers' expectations. Such a commitment is an important driver of sustained growth.

- **Evidence-based decision-making**: Reliable data analysis and informed, strategic decision making enhance the chances of achieving desired outcomes.
- **Relationship management**: The process of building strong, clear, and mutually beneficial relationships with all stakeholders and interested parties lays the foundations for sustained success.

Anecdote: After obtaining my undergraduate degree in Computer Engineering and working for a couple years running a small team, I had an opportunity to return to obtain my master's degree from the University of Virginia's Darden School of Business. It was quite an experience and I still remember the 100-case party, an event of significance as it meant we completed 100 case studies in our first term. At the time, UVA did not accept direct admit students, however, after working for several years, I came to understand why. It is easier to obtain knowledge if you have context. As a lifelong learner, I was excited to dive into management issues, and to understand management dynamics across various competencies, industries, and situations. Participating in case studies that span marketing, operations, human resources, ethics, sales, and finance was eye opening. Quality management, in particular was one of those topics I naturally gravitated to since I am a person who is interested in process improvement and seeking ways to make things better. This would lead me to achieve one of my career goals of becoming a consultant at one of the big six. One of the interesting books that was part of our instruction was 'The Goal' by Eliyahu M. Goldratt and Jeff Cox. Written as a novel, it is a story about the process of ongoing improvement. In Chapter 4, we get to a definition of productivity. "Every action that brings a company closer to its goal is productive. Every action that does not bring a company closer to its goal is not productive." Later in the book, we learn about Herbie, a boy scout who feels bad because he slows down the troop while hiking. The realization came as the troop tried to figure out how to get Herbie to go faster to achieve their goal (hiking to Devil's Gulch).

This goal was accomplished by a four-step approach. 1. Admit that Herbie is the slowest (identify the constraints), 2. Discuss why he is so slow (understand the constraints), 3. Help Herbie to go faster (identify options and plan of action), 4. Implement the solution and get it done (execute plan). In retrospect, learning about the standards developed by the experts and reading leading business books like The Goal, provided part of the foundation of being a successful consultant. One of the lessons I personally learned was to first understand processes by actively listening to those who know the process best. Identifying constraints and figuring out the significance of each constraint, allows you to create a plan of action that provides the greatest benefit/improvement. Evidence-based decision making has driven me to focus on data-driven solutions. Many of the experts I have worked with in this field are hyper focused on understanding what is the goal of the organization and then aligning the key performance indicators to align to those goals.

Apart from educating yourself to provide you with the foundation, you may ask, "Where do I begin in my journey as related to quality management?" A good place to start is to review ISO 9001 (ISO 9001:2015, 2015). The ISO 9001 requirements "define how to establish, implement, maintain, and continually improve a quality management system (QMS)". Another resource is the CMMI Institute (ISACA, 2024). Their appraisal is an activity that helps you to identify the strengths and weaknesses of your organization's processes and to examine how closely the processes relate to CMMI's four capabilities and six maturity levels. The three objectives of an appraisal are:

- Determine how well the organization's processes compare to CMMI best practices and identify areas where improvement can be made.
- Inform external customers and suppliers about how well the organization's processes compare to CMMI best practices.
- Meet customer contractual requirements.

Tactically, I have worked on teams with resources focused on QMS. The specific steps followed include:

1. Establish business needs and performance objectives - What is the goal?
2. Perform an internal QMS audit (BECNT, 2023) our appraisal - Understand the constraints.
3. Identify Goals, Questions, Measures, Targets, Thresholds and Corrective Actions - Identify options and processes that allow you to navigate your roadmap with contingencies.
4. Execute - Start the plan, periodically reviewing and adjusting as needed.

To facilitate this process, one simplified technique is provided below:

- Identify **keywords** from the Vision, Mission, and Objective statements
- Map the keywords to quantitative **attributes**
- Define, select, and prioritize associated **measures**
- Map the measures to **measurement sources**
- Create or update **quality management policies** including the **process** for continuous monitoring and periodic review

7.3.3 Supporting the Customer

After a solution is developed and ready for rollout to the user base, it is important to have a training and communication roll out plan. Again, for business intelligence and data analytic teams, standardizing helps the team to be able to grow quickly.

7.3.3.1 Training (Internal/External)

Training is one of the most important parts of establishing a highly performing team. At least one person on the team should be a strong facilitator and have competencies like patience, understanding your audience, and leveraging different ways of instruction to ensure the lessons being taught are well received. BI teams often have two primary approaches to training. The first, and the one most often used for large customer bases is 'a train the trainer' model. In this model, the team trains up to seven trainers with new capabilities, features, reports, or navigation changes focused on the requested solutions. The 'train the trainer' model usually requires someone who is familiar with the business process and either developed or has a strong understanding of the solution based on the requirements. The second approach is to have a facilitator who can instruct new users. The difference between these two types of instructions cannot be overstated. Instructing all end users may seem to be intuitive; however, once you get to over fifteen individuals in a class, there are other things to consider.

- Will you need individuals to help assist during the instruction?
- Do the participants have access to the solution-training environment?
- Are the users familiar with the tool the solution was developed in?

- Do the participants understand how the data is sourced and stored?
- Do they understand the business process?

Having strong facilitation skills is essential for this type of training. Most important is the collaboration amongst the functional area, business intelligence team, and the data stewards of the related source systems. All three areas should be active participants in any business intelligence or data analytic solution training. Most business intelligence solutions support a defined business process. The functional area should always drive any training with the end users. This team understands the business process in detail and how the data enters the business and how the solutions created by the BI team are used to support the business process. Source data stewards should next provide an overview of the key data elements captured and stored in the system. Understanding the order of operations, key data elements, and the life cycle of data helps to set the stage for how the data can be used to make informed decisions in their specific role. Finally, the Business Intelligence team, working with the functional area, can help create step-by-step workshops that are specific to demonstrating navigation and capability of the solution that are most pertinent to the individuals attending the session.

7.3.3.2 Supporting the Customer in PROD

In a DEVSECOPS or SECDEVOPS model, one of the roles of the business intelligence team is to support the end user in a Tier two or Tier three capacity. *See Section 6.2 Customer Support Monitoring* for more information on Tiers. It is important to create an expandable model as early as possible in the team life cycle. Planning for growth is essential as a team can quickly get overwhelmed with requests if they are good at providing solutions that the customers see as valuable. As the user base expands, the BI team is more challenged to create solutions more quickly for a more diverse user base. Some teams allow for a self-service approach, which gives the functional areas additional capabilities like creating their own reports or query capability for conducting their own data analytics. This approach is fine; however, there are other considerations that the management needs to review to customize the implementation of this approach. What are the competencies of those who will be granted this capability? What type of capability will be provided? Is the infrastructure for the team isolated (i.e., do they have their own servers or share with other teams)? If it is determined that a self-service or semi-self-service model is warranted, a standards guide should be developed to confirm basic competency in working within the environment. This includes rules or guidelines that should be followed. Something as simple as having a code review with one of the DBAs or SMEs on the business intelligence tool is important to minimize the risk of other teams in a shared infrastructure model. Even if the self-service infrastructure is isolated, having review gates is helpful to ensure that code or solutions being generated does not affect the larger team due to inefficient query development.

7.3.4 Measuring the Team's Effectiveness

Measuring the team's effectiveness allows you to justify support level of the team and quantify some of the value the team provides. As part of a management team, it is important to continue assessing findings from key performance indicators and supplement with subjective factors to provide a good assessment of the current team. Once again, understanding the organization's mission and aligning the metrics and subjective criteria of each of the team in a model of continuous improvement helps to achieve a team that is continuously learning and growing, optimizing their high-performance status.

7.3.4.1 Development Metrics

Development teams who adopt an Agile development life cycle with complimentary tools have numerous resources available to help in measuring and there have been metrics that are known and easier to track if implemented correctly. One of the most well-known is the Sprint team's velocity. This is the amount of work the team produces over a Sprint. Although it may sound straight forward, there are still a number of factors that may lead to better measurement. Understanding how to create effective story points, ensuring team members are entering time into subtasks on at least a daily basis, and sharing improvement opportunities via retrospectives are areas of maturity that lean to continuous improvement and higher performing development teams. Other aspects of measurement for visualization may include the type of work being performed, and who is performing the work by role or individual.

7.3.4.2 Security Vulnerability Remediation

Dashboards can be created for the security team to monitor and measure the number of vulnerabilities identified by each security-monitoring tool across each server. Information about vulnerabilities like applications that use the code, which servers the vulnerability resides on, which products are affected, and criticality of the vulnerability should be captured. Monitoring these metrics over a period of time allows management to allocate resources and manage the remediation activity by prioritizing based on the criticality of the vulnerability.

7.3.4.3 Infrastructure Team

From an infrastructure perspective, support tickets can be measured by role (i.e., Database Administrator, System Administrator, Release Manager, etc.), type of request (i.e., index creation, query review, performance optimization, new report support, etc.), and how long it takes to complete requests. Again, these can be measured and monitored over time to identify bottlenecks in the process in order to make better decisions on resource allocation and task prioritization.

7.3.4.4 Customer Service

There are at least four metrics that any service organization with a ticketing system for customer support can use to measure the team's effectiveness.

- Average Request Remediation Time: The length of time it takes to remediate customer service requests. The goal is to continuously seek to reduce the time it takes to resolved issues.
- Average Number of Closed Tickets by Tier: For each tier (1-3), identify the number of tickets resolved. The goal is to resolve as many requests as possible with the lowest tier.
- Number of Tickets by Category: Once you defined the best categories for your support request, the goal is understanding which category should be focused for remediation through training or updated support scripts

- Number of Tickets by Role: Understanding the roles of the submitters may allow for more rapid remediation by implementing a targeted training plan.

8 DATA MIGRATION PROJECTS

The objective of the Data Migration Projects section is to focus on optimizing high-performance teams that are involved in modernization efforts (i.e. moving from one technical platform to another).

Primary Goals:

- Unique Capture Requirements
- Data Migration Life Cycle Components
- Shift to Agile

8.1 Data Migration

The complexities and the quality of an organization's underlying data that it curates and manages is often an area that is frequently underestimated. Data migration should be a separate initiative that runs in parallel with system development and it should begin early in a new system life cycle. The timeline for completing the initiative is directly proportional to the length of time the system being replaced has been in existence. Other factors that affect the complexity and timeline are the maturity of the organization's governance process. Throughout the initiative, a review of the data will likely drive new requirements. Specific tasks that can be accomplished include:

- Capture all reference data usually identified by the valid values of application drop down selection elements.
- Complete a mapping document on both the field and reference data fields
    - This is most critical as it will capture all business rules for transforming data
- Determine the process for updating and approving the final element names, valid values, and definitions of both by the data owners
- Review and confirm the Record Information Management Schedule. For older systems, try to limit the amount of data you are migrating to only what is necessary.

Data migration projects can be complex in nature, and having data subject matter experts knowledgeable in these aspects of data is important to achieve success. If the organization does not have a mature governance process that includes Records Information Management (RIM), it is highly recommended that an archivist role be assigned to someone early on the project. Having a clear understanding of the types of data and a clearly defined records schedule allows the program to predetermine what is the minimum data required for the migration. You can begin or continue your growth in this area by exploring the information provided by the National Archives and Records Administration (NARA). NARA is the United States' record keeper.

> **Note**: Only one to three percent of records used for business conducted by the United States federal government are deemed important for legal or historical reasons.

Samples of Record Control Schedules are explained and provided on their site. Reviewing and understanding the purpose and objective of good record management allows you to determine what is best for your organization. International organizations such as the Council of Europe publish related documents such as the User Guide to Retention and Disposal Schedules. Other international associations focused on records managers and administrators such as the ARMA International Council on Archives Council of State archivist (ARMA, 2024), are good to review to determine and optimize the depth around this topic. Finally, if time is of the essence, one additional alternative is to look for any upcoming conferences for record managers to build your network of individuals passionate about this area and to quickly learn about current challenges and how they are being addressed from an industry perspective.

8.1.1 Data Related Projects - Data Migration

Standardization benefits any type of project from the simplest to the most complex. Having defined processes and a consistent way of performing work helps with expediting the solution delivery in a way that produces high quality products. One example of a complex project type that has many interdependent components is the work on data migration projects. This is particularly true if the data migration activity is occurring from a legacy system that has been around for longer than ten years to a new onsite or cloud-based infrastructure. Many of the cloud vendors have tools that help to make the transition easier. However, if the organization has standardized tools, processes, and governance in place, it helps to make any data migration activity smoother.

- Does your solution capture and implement business rules on data received by other systems of records?
    - If so, it is important to clearly understand who to go to provide instructions
    - Profile source data including key fields that require values; field sizes, special character usage, and the quality of fields used in business rules. Providing these up front minimizes issues during migration.
    - Acceptable downtime for migrating active data between systems - architect as appropriate
    - Prepare automated routines to confirm transition of data which may include simple row counts or summing select numeric fields to ensure fidelity
    - Confirm that the backup and restore procedures are working effectively

- Is your solution the source system of record (SOR) for any of the information collected?

Any organization undertaking any type of data migration effort should create a playbook that stakeholders can use to clearly capture the key components to achieve success. The playbook allows the various stakeholders to better understand the big picture and their role in making the initiative a success.

8.1.1.1 Architecture Overview and Mapping

Consistent communication is very important for data migration programs. Two of the most critical roles are the solution and data architect roles. The larger and more complex your organization becomes, having team members who are knowledgeable in these competencies who can quickly expand these areas for your organization will be more valuable. If your organization is expected to grow in the near future, having a consistent way of communicating your products and information assets will save much time in the future. Strong individuals in this area are competent enough to facilitate and capture clear business requirements from the business. They should be able to work with the technical team to determine clear technical requirements for the technical team. With the help of solution architects, technical leads, and others on the management team, they help set the direction for what is expected and how it will be accomplished. As individual system developers, testers, and other team members fulfill their duties, clarity in these two areas helps to drive business and technical questions early in the process, which avoids building unnecessary technical depth. Important information to capture are:

- Source System Information

- Target Solution
- Architecture/Feature mapping from source to target

8.1.1.2 Governance

One certainty in any large data migration effort is that the team will run into unexpected issues with the data. Considering the governance of any data migration initiatives allows the organization to minimize the risk by building or expanding processes to clearly identify how issues will be resolved including ownership and turnaround time for resolution. On large programs that expand across divisions or organizations, it may be setting up a governance council with the appropriate level of team members who have authority to make decisions for the critical division. This may be one council or split into multiple councils depending on the types of functional or technical issues. The information captured as part of the governance register includes:

- Source Data Owners
- Process for source corrections
- Target Data Owner
- Reference Data POCs
- Downstream Customer POCs
- Resolution process across each via the governance council

8.1.1.3 Order of Operations

Understanding the order of operations or workflow of new or changed functionality of a solution is important in efficiently delivering a viable solution that meets the user needs. When the team experiences unknown situations and need to explore options to provide to the business owner for remediation, the criticality of this knowledge can be utilized to make better decisions for data migration projects. In understanding the workflow, the team should explore a day in the life of a data element. This includes the full life cycle of data from curation through transformation and dissemination. Understanding this workflow allows the team to determine the best way to sequence the data for migration, which items should be confirmed by the source, and what is the best process for migrating the data. Having a clear understanding of the workflow that includes dependencies during the process, upstream and downstream, will lead to more successful and efficient migration. There may also be different levels of data movement that are important to capture (i.e., administration maintenance vs user input for curation). Having a good understanding of the way data is curated and changed up front helps to create a plan and, along with the governance processes, will allow the team to capture a majority of issues the teams typically experience in data migration.

8.1.1.4 Transition Methods

Understanding the various options of migrating data across environments is important even if you only utilize one standard method of transfer. By documenting options upfront, it allows you to quickly mitigate or find alternative solutions in the event one is needed when an unknown situation pops up during the migration process. It is much easier and less stressful to determine options upfront which also helps with the level of stress during a crisis of an unknown event. Information identified and captured in the transition methods can be used to populate other deliverables like the Interface Control Document (ICD), which is the deliverable that captures key information related to upstream and downstream data exchange and methods. At minimum, there are five pieces of information required for transition method information capture, which includes the data structure and flow, the format, key fields for internal audit, the internal quality assurance process, and the names of the maintenance points of contact.

The first area, structure and flow identify the various options that could be used to transfer the data along with the rationale of each and the recommended method that will be standard. For example, is the migration going to be done over the network, backed up and restored via an appliance, or will a third party perform the service? The method or process for the migration activity should also be captured. The Order of Operations may be helpful in determining the best flow for the data migration activity.

Key fields for internal validation and verification on the integrity of the data migration initiative are critical for acceptance testing and adoption of the newly migrated data by the user community. Reports will need to be quickly generated to support any perceived discrepancies and to build a level of confidence for a successful migration. Organizations often find insight into the quality of underlying data and use the opportunity to make changes to better suit the organizational goals.

The verification process of data migration is important to the integrity of the data migration initiative. If there are questions on the data transfer process, sponsors and customers of the new solution that the data is migrated to will be resistant to accept the results. Customer representatives familiar with the data and producing reports on the data set make good candidates for a temporary validation and verification role. These resources inherently understand the expected value of reporting elements and causes for inconsistencies. A select number of users like these, who are trusted by customers using the new system, should be part of the validation process in order to mitigate adoption risk of the new solution.

Finally, it is critical to understand the primary point of contacts responsible for resolving technical issues during the migration process. Data stewards who manage and maintain the source data files as well as those responsible for validating the underlying data should be documented as part of a stakeholder registry. The full list of names should be confirmed the week before the migration activity begins. This is particularly important if the data migration is transitioning to a new transactional system where downtime must be kept to a minimum. Instances like this usually require a well-coordinated tight schedule of activities after core hours. Understanding the process of quickly remediating issues by everyone on the team is key to a successful project.

Anecdote: One leading practice tip is to create a conference call line for the duration of the physical data migration so the team can speed dial if and when they need to interact across team or individual handoffs. Other tools like a group chat or instant messaging can also be effective.

Having a consistent way of communicating across various stakeholders becomes even more important if there are multiple parties involved in the data migration.

A clear understanding of the key fields that changed, the lineage of the key fields, the process used to validate, and the point of contact for specific questions on the transfer allows the business to educate/enhance the business use of data and keeps the technical team focused on effectively implementing the process as designed. If design changes are required, key decision makers can be quickly brought together to assess the impact of any changes, to prioritize, and to quickly adjust requirements as needed. Using the approach outlined above will allow the team to be flexible and quickly react to the unknowns that result in data migration initiatives.

8.1.1.5 Data Elements Mapping

The Data Element mapping template is the primary artifact used in any data migration project. As a result, one can get a good understanding of the organization's data maturity and likely success of the project by reviewing the type of information captured in this artifact. This document answers the basic questions of the data migration such as where is the data coming from? In addition, where is the data migrating? How the team captures and communicates information about the data provides insight into how well the technical team communicates as they build the migration solution. A list of source elements is captured with characteristics including and not limited to the type, the size, constraint, default values, and reference values as applicable. Reference data with matching criteria are captured as needed. Similarly, the data characteristics of the target system are captured with associated characteristics similar to the source elements. Finally, the source data element to target data element mapping is completed, which includes the transformation business rules and any exception processing.

8.1.1.6 Performance Management

Performance Management is an area that covers two types of performance. First is the speed it takes to migrate the production data. If the program has budgetary constraints, the duration of a production migration of an active system is critical because the customer may lose capability for a period during the migration process. The source data is profiled to get a better understanding of how much data is being migrated, the response time on a sample set of existing queries, and the optimization techniques used during the process. Having a strong data architect familiar with the technology being used is most important in understanding the performance management requirements.

Secondly, performance management is a tool used to capture the baseline and measure the new system's response time. Comparisons are often used to understand the performance enhancements of the new system, which is often one criterion for acceptance. Information usually captured in this area includes identifying which fields should be indexed to enhance data load and retrieval performance. The load strategy, comparing the speed of resulting queries with those captured as part of the baseline, determining remediation steps for mitigation of performance hurdles the team may run into including long running queries and identifying and monitoring queries that pull back large amounts of data. A knowledgeable Business Analyst should also capture the performance of the existing system to baseline the performance of the new system.

8.1.1.7 Checklists

Due to the complexity of large-scale data migration projects, it is best to prepare checklists to minimize risk and increase the chances of success. When combined with formal checkpoints leading up to the actual data migration, these checklists can be an effective form of tracking and communicating completeness and status throughout the project life cycle. This is particularly important for large programs that rely on external parties for the source data being migrated. A source preparation checklist is the most critical of these. This checklist should include those items required from the source, including any insights from the data profiling activities. The insights may include cleaning the existing transactional or reference data required prior to the migration effort; completing the data mapping document with required fields and transformation logic; and documenting of the technical, functional, and governance points of contact prepared to support the effort.

From a technical perspective, the checklist may include information on the data migration process; points of contacts, confirmation of collaboration tools to be used during each step of the data migration, and the verification techniques that will be used by the technical team. It is most important to plan and communicate what the team should do if specific issues arise. For example, a conference call number may be set up so that the team periodically meets on the same line when transitioning activities during migration, or when an issue arises that require multiple parties. This one pre-planning step alone will mitigate risks, help to achieve success and enhance team building across core technical team members responsible for a successful migration.

8.2 Data Related Projects - Shift to Agile

Companies continuously explore ways to provide value added products and services to their customers. Either those companies who are not yet exploring new software development techniques will start doing so soon or they will become obsolete. As companies look to shift development life cycle methodologies from conventional methods used in their legacy systems, the need for setting up the correct structure and implementing a shift in culture to make the change becomes increasingly more important. Understanding the tools and processes to make a successful shift helps to enable the change to new expedited life cycle techniques. In this section, we will explore some of these tools and techniques to help navigate such change.

8.2.1 Agile Supporting Tools

Agile is a methodology used for expediting a system development life cycle. There are twelve tenets based on the Agile Manifesto, which are listed below for reference.

1	Our highest priority is to satisfy the customer through **early and continuous delivery** of valuable software.	7	**Working software** is the primary measure of progress.
2	**Welcome changing requirements**, even late in development. Agile processes harness change for the customer's competitive advantage.	8	Agile processes promote sustainable development. The sponsors, developers, and users should be able to maintain a **constant pace indefinitely**. (i.e., Rhythm)
3	**Deliver working software frequently**, from a couple of weeks to a couple of months, with a preference to the shorter timescale.	9	Continuous attention to **technical excellence and good design** enhances agility. (i.e., QA and Risk Management)
4	Business people and developers must work together daily throughout the project. **Collaborate**	10	Simplicity–the art of **maximizing the amount of work not done**–is essential.
5	Build projects around motivated individuals. Give them the **environment and support** they need and trust them to get the job done.	11	The best architecture, requirements, and designs emerge from **self-organizing teams**.
6	The most efficient and effective method of conveying information to and within a development team is **face-to-face conversation**. (or remote video session)	12	At regular intervals, the team **reflects on how to become more effective**, then tunes and adjusts its behavior accordingly. (i.e., **Retrospective**)

Table 6: Agile Manifesto Tenants

Agile implementations can be significantly simplified by planning and using the correct tools early in the process. It is important to understand which tools will be used to plan and track tickets; code, build, and release capability, monitor security and identity, and collaborate.

Plan and track development tickets: Having a standard tool to organize your development work is one of the most essential tools for Agile projects. The tool should have the capability to capture epic user stories, decompose the epic to workable user stories, plan releases and sprints, and distribute tasks across the project team. The ability to create quick filters and provide flexibility on reporting of the work in progress is an important feature. The tool should easily allow work to be prioritized and discussed holistically as part of your release and sprint planning exercise. *See Section 5.2.2 Sprint Planning* for additional information. More mature tools in this space offer flexibility in creating workflows for managing governance, release tracking, and reporting.

Code, Build, Release Capability: A second important tool to take full advantage of rapid development is a version control system. The primary component is the repository where code created on a local system can be held in one location and versioned to keep the related files together. In many cases, the development team uses terms like Version Control System (VCS), Trunk-based development (TBD) and branching, for new updates in development. This capability allows for creating different software components in parallel and allows the development team to enhance the quality of the code base by isolating changes without affecting the trunk.

Testing Tools: A third tool used to expedite a rapid development environment is the testing solution. Selecting a tool that is flexible to perform both manual and automation testing, capturing and summarizing results, and integrating with other enterprise standard tools will provide the development team with the flexibility they need to get their work done efficiently and increase the quality of the code developed. Metrics, such as defect density, can easily be defined and implemented to monitor the root cause of failure for continuous improvement of the process. Automation allows for more rapid regression testing of the entire code base to help with rapid root cause, impact analysis and remediation by the development team. More mature teams include testing for invoking testing on multiple types of infrastructure, such as service-oriented architecture (SOA), representative state transfer (REST), web services, the capability for load and performance testing, and reporting the results across the enterprise solution.

Collaboration Tools: Collaboration solutions are critical for cross-team communication. These solutions are even more critical if the team is not in one location. Having tools that allow quicker decision making by bringing the team together is critical to take advantage of rapid development techniques. This solution typically includes a combination of knowledge management, messaging, and reporting. More mature solutions include the capability for making voice and video calls, and automated workflows. More integration with other agile and enterprise tools will provide greater flexibility to optimize solutions that best serve the organization.

Monitor Security and Identity Management: Security and Identity management become increasingly important for any organization today. Having tools that allow for single sign on (SSO), the ability to quickly audit, automate user lifecycle management, authentication, token control, active directory synchronization and security integration are essential components. This area becomes even more critical as organizations move to the cloud. Tools in this area should allow the team to optimize robust and scalable governance plans while providing the flexibility and collaboration required for rapid development. As many organizations select a hybrid solution, the complexity of the security and identity management solution increases and requires tools that can support the increased architectural complexity.

Cost is one of the biggest factors in the tool selection process. As organizations move to rapid development and migrate to new technologies like the cloud it is important to do an analysis of alternatives (AoA) in each of these areas to determine which products work best in your environment. An inventory of existing software and a roadmap to get to your ideal solution mix is important. Flexibility and integration across multiple products and infrastructure will allow for the agility that may be required to avoid locking into proprietary solutions and creating the ideal mix of tools for your team to truly be agile.

8.2.2 Backlog Grooming

A backlog is best defined as the list of work items not yet completed. The tenets most closely related to effective backlog grooming are numbers one, two, six, and twelve in Table 6. The concept of backlog grooming in Agile provides the team with an opportunity to interact more with the customer and build a relationship through effective time management and meeting facilitation. Effective backlog grooming sessions start out by having a well-defined and usable backlog. A backlog should, at minimum, report on the following attributes to have successful and engaging discussions.

- **Priority**: Identifies the priority of the Epic or User Story requirement. Most tools have a standard reference list such as Critical, High, Medium, and Low. Ideally, these items would be force ranked to help in determining the most valuable work to develop at any given time.

- **Requested by Date**: An optional field that captures when the business requires the capability in a standard date format.
- **Requested by Date Rationale**: An optional field that captures business reason for the need by date
- **Ticket Type**: All work being performed by the team should be listed in the backlog. In many cases, there are different types of tickets the team should track for reporting purposes. These may include and are not limited to New Requirement, Enhancement, Defect, and Maintenance.
- **Change Type**: Another field that can optionally be tracked includes the type of change. Changes may occur in the code base, scripts, analytics, and infrastructure. Sometimes teams combine the ticket type and change time, based on how the team performs the work.
- **Ticket Number**: The ticket number is a unique identifier for each item in the backlog. This number may be alphanumeric to help group or categories tickets. The ticket-numbering scheme should be clearly understood by the team, as it is the primary way to quickly communicate a specific requirement under discussion.
- **Ticket Summary** - Having a title or a brief summary of each item in the backlog helps the team understand the topic and determine if it is the correct topic to review or update. Having a clear standard on how ticket titles and summaries are written saves a lot of time when it comes down to execution.
- **Status** - The status field is used to indicate where the ticket is in the ticket life cycle. It may be as simple as Not Started, In Progress, and Complete, to more complex status that includes workflow components.
- **Release**: The release field is used to capture the release number once assigned. More mature programs have multiple fields to

easily track planned versus actual reporting or to designate where in the life cycle a ticket is requested. For the initial grooming sessions, discussions on a particular ticket, the 'target release' is usually the focus.

- **Sprint**: Similar to the release field, the sprint field may be captured to help with sprint planning. Multiple fields may be used to capture changes in sprint over the life cycle of implementing the work item. Populating the field is usually optional and changes over time.

- **Technical Complexity**: The technical complexity field is very important in sprint planning and in communicating the work that is being executed by the development team. The development lead usually is the person who scopes and determines the complexity of the associated work effort. It is usually measured in terms of High, Medium, and Low. Teams that have clearly defined standards of complexity levels help in cross-team communication. Product owners review this field as part of the cost benefit analysis used in setting their priority.

- **Technical Complexity Rationale**: The Technical Complexity Rationale field is used to give additional context and commentary on the technical complexity rating. This is often completed by the development lead who performs analysis on the work requested.

- **Story Points**: The story point's field is used to capture the relative size of effort on the user story. Teams may start out using effort hours or the relative sizing of the effort compared to other known activity usually performed by the team (i.e., extra-large, large, medium, and small). More mature teams use more complex methods like assigning Fibonacci numbers. It is important that, whichever method is used, it is clearly understood by the team. This field is a primary field used for sprint planning.

The business analyst is usually the most appropriate role to facilitate these discussions and to work directly with the product owner on updating and maintaining a healthy backlog. Scheduling recurring meetings with the product owner for continuous grooming of the backlog provides enhanced communication across the team and leads to a more productive relationship between the customer and development communities. Customers understand that they have an opportunity to change the development of work items up to the time the work product is assigned to a sprint. If sprints are kept at two-week intervals, it provides for better discussion on tradeoffs for which work can be replaced. It also provides the development team with flexibility on how best to implement or modify code changes. There are two typical scenarios for loading a backlog with change requests.

Scenario 1: Product owner has many change requests not yet entered in the backlog

In the first scenario, teams usually either are new to the agile process or have a separate requirement approval process that has not incorporated the creation of tickets in the backlog. Starting out with a backlog, it is recommended that you focus on no more than two release cycles when creating and grooming or decomposing your requirements into work items. If the team has a robust application where you can filter based on the targeted releases, a dashboard or report should be used to facilitate the recurring backlog grooming sessions. Keeping these sessions separate from the execution of the work has benefits including the frequency of such sessions. It also allows the product owner flexibility in planning how to communicate upcoming capabilities to the customers and how to secure the appropriate funding from the business sponsor.

Scenario 2: Product owner enters change requests into the backlog as received

Utilizing a tool that integrates requirement capture with a backlog helps in reducing complexity and communication of which requirements will be implemented and the timing of seeing the change. If all change requests are entered into the backlog as part of the requirement approval process, it facilitates the process by consolidating the primary work the team needs to achieve.

For both scenarios, it is important that any requirement be given priority and an expected need by date with rationale. As tickets are created to decompose new requirements and enhancements, they should reference the associated requirements. Whenever possible, the team should try to decompose all tickets to a level that can be accomplished within one sprint. Scale Agile takes this one step further and defines several different types of backlogs. *See Section* 6.2.1 Grooming and Pruning for more details. Each ticket should have clear acceptance criteria, dependencies, and constraints. Having a repeatable and effective grooming process allows the team to establish a foundation to perform better release and sprint planning which we will cover in the next **Section,** *8.2.3 Release and Sprint Planning.*

8.2.3 Release and Sprint Planning

Once baseline requirements are identified, prioritized, and decomposed to deliverable capability, it makes it easier for the technical team to rapidly develop the capability in the most efficient manner. This creation of epic user stories may be sufficient for planning major capability for a release and decomposing the epic user stories into more manageable pieces of work and will be required for the most efficient development process. Using rapid and iterative development techniques like Agile means that it is likely the requirements will be refined until the decomposed user stories make it into a sprint. Approval for work that goes into release and sprint usually falls under multiple roles. From a release perspective, the primary sponsor, the product owner, and the scrum master are the primary decision makers, and they establish the expectations and commitment of the development team. For sprint planning, the roles that typically finalize scope include the scrum master, and the development leads.

Having a mature backlog with well-defined prioritized requirement user stories facilitates good release and sprint planning. It is easiest to use a spreadsheet tool to create a process for quick, iterative, and effective review and planning of releases. Most of the tools used for the backlog repository have the ability to export in a comma-delimited format, aka comma separated values (csv), that can be imported to your spreadsheet of choice.

First, create a table with the fields that are most beneficial for the exercise. At minimum, this should include Business Value, Technical Complexity, Ticket Number, Title, and Story Points. Fields like software components may also be necessary depending on the size, complexity and mix of the technical team. If there are multiple components being developed by the same team, it is important to understand any expectations on the percentage of development required for each component. Organizations usually have different funding streams and aligning these expectations up front helps in planning out the work.

> **Note**: If one sponsor provides three fourths of the funding for the team, approximately seventy five percent of each release should be on delivering capability for that sponsor's software component.

The second step is to sort your table by business value, highest to lowest, and technical complexity, lowest to highest. For teams who are already well established in Agile and have a solid understanding of their rhythm and velocity, the exercise becomes a bit easier. For those teams who are still maturing, conservative estimates should be made during the initial estimating stage and the high-level effort hours can be used for the exercise. Many teams only do this exercise for new capability and enhancement to their existing capability. It is important that all known work, including known technical changes, be also incorporated for effective planning. Each sprint should be assigned with no more than eighty percent of the team's availability initially. Other detractors of work should be incorporated up front to create a healthy development environment that is sustainable for long-term rapid development. As a project sponsor, will I accept that I am only paying for eighty percent of the development time? Yes, if the product sponsor sees results, has an open line of communication, and understands what is being used with the other twenty percent? No reasonable sponsor will object to this standard. So, you may be asking, what is in the other twenty percent? The remaining twenty percent includes scheduled times for meetings, such as grooming sessions, retrospective, collaborative, testing/remediation, configuration management, and operations support, emergency releases, drive-bys such as small urgent action required by the sponsor, and time for research and root cause analysis.

Finally, review with the team to provide an opportunity for feedback and commitment to the plan. The backlog tool can be used to create a dashboard of the proposed work for the upcoming sprints strictly based on the logic method described in the second step above. Meeting with the team allows for additional changes that may be more logical depending on the architecture and code development components. The team may want to swap items that make more sense technically.

Using this approach results in creating an efficient list that can be communicated with the broader team. There are two points to add here. First, because the sprint items are logically completed by order of value and technical complexity, it inherently helps to better communicate the most valuable components to the team as they are making decisions throughout the development life cycle. Second, the sprint and overall product backlog is still available and the product backlog is consistently being updated and reprioritized. The development team can pull items from the list, as needed, if they have extra time allocated. From a project management perspective, providing supporting tools and processes that ensure high quality development code and facilitation techniques to minimize unnecessary meetings helps the team to achieve great things. It is also important to establish rules on when changes can and cannot be introduced. For example, once a sprint has started, it is highly recommended that no changes to that sprint be introduced. Making changes to a sprint after starting directly affects the team's velocity. If the sprint is two to four weeks in duration, and this occurs too frequently, you may need to adjust the size of your sprints. The disruption caused by introducing requirement changes in the middle of a sprint disrupts the rhythm and achieving the maximum team velocity.

Once you have planned your upcoming sprints and the team commits, the project management team can focus on supporting the development team by determining the best tools that can continue to improve the development process, communicating the progress, and facilitating effective meetings. The most important meetings to establish the rhythm of the team using rapid development techniques are the daily scrum, scrum of scrums, and retrospective meetings.

8.2.4 Scrum and Scrum of Scrums

Scrum is a set of practices used in agile project management that emphasize daily communication and flexible assessment of plans that are carried out in short, iterative phases of work. One of the most critical meetings to facilitate using agile development is the daily scrum meeting. During this meeting, everyone on the team states the work accomplished since the last meeting, the work anticipated until the next meeting and indicates any specific roadblocks. The focus is to inspect the progress toward the commitments made by the team on the sprint in progress. Most often, a scrum master or project manager is used to facilitate the discussion, ensuring the primary code development team members actively participate in the discussion. Other supporting members may listen in or be available to discuss roadblocks like environment availability or any issues or clarification on the solutions. The meeting should be held at the same time every day and is often held to a maximum of 15 minutes. Some teams literally stand for the duration of the meeting to reemphasize this point. A good facilitator that starts on time and methodologically goes through the key members saves time.

On larger development teams, several scrum teams may be created for each solution subcomponent. The scrum master on these teams usually get together for a separate Scrum of Scrum call that brings together any pertinent topics required for cross-team communication. This meeting is also typically 15 minutes. The result from a project management perspective is that it allows and enhances daily communication and allows the team to schedule effective meetings for the purpose of removing roadblocks. Thus, leading to a more efficient development process that can continue to be proactively improved through the retrospective meeting.

8.2.5 Sprint Review

Sprint review sessions are a great way to enhance communication with the product owner. The session is usually conducted at the end of each sprint, and it is a means of presenting the completed work to the customer to gather feedback on the completed work. These meetings usually lead to additional refinement and guidance in future development work and sprint planning. These sessions typically are scheduled for 30-60 minutes depending on what is completed for each sprint. It is highly recommended that the session is recorded to capture feedback on the code being reviewed to allow anyone on the development team to digest and replay direct customer feedback on their code.

8.2.6 Retrospectives

Retrospective meetings typically reflect the maturity of a team. Project managers who are new to an existing team can learn a lot based on the participation level of the team. Team members relatively new to agile typically just want to code and perceive meetings such as this one as administrative and overhead. If not facilitated well, it increases the likelihood of repeating mistakes of the past. On mature teams, participants actively participate, document, and prepare a log with action and ownership for lessons learned items. These open action items are reviewed during future sessions until the team agrees that it can be closed. This session usually reflects on the past sprint and asks three questions: what went well, what did not go as well as expected, and what could the team do to improve in future sprints. More mature and larger teams also categorize each entry to suggest if the change is only related to the project, the program, or the organization. For those suggestions outside the project, the team only provides recommendations; the scrum master can take those to the appropriate group for assessment and inclusion.

8.2.7 Hurdles/Opportunities

Any organization considering rapid development techniques like Agile should do so as soon as possible. There are challenges, however, the benefits, in the long run far outweigh the short-term challenges of the decision. Challenges include balancing the amount of process and supporting documentation changes with workable code, understanding if you should stop inflight initiatives and roll out the new processes immediately, culture shock, or more gradually, culture change. Engaging the right resources to affect the change best suited for the organization is important. The Agile methodology provides great potential and supports the scaling of high-performance teams. Gaining an understanding of the approaches used in Agile is important as you work with this methodology. Two references I have found helpful in this regard are provided in the two tables below.

	SCRUM	KANBAN
Release methodology	Regular fixed-length sprints of two to four weeks.	Continuous flow
Roles	Product owner, scrum master, development team	Continuous delivery or at the team's discretion
Key metrics	Velocity	Cycle time
Change philosophy	Teams should strive not to change the sprint forecast during the sprint. Doing so compromises learning around estimation.	Change can happen at any time

Table 7: Scrum vs Kanban (1 of 2) (Atlassian, 2024)

	SCRUM	KANBAN
Origin	Software development	Lean manufacturing
Ideology	Learn through experiences, self-organize and prioritize, and reflect on wins and losses to continuously improve.	Use visuals to improve work-in-progress
Cadence	Regular, fixed-length sprints of two to four weeks	Continuous flow
Practices	Sprint planning, sprint, daily scrum, sprint review, sprint retrospective	Visualize the flow of work, limit work-in-progress, manage flow, incorporate feedback loops
Roles	Product owner, scrum master, development team	No required roles

Table 8: Scrum vs Kanban (Atlassian, 2024) (2 of 2)

Understanding the differences and which approach is best for your team helps in agile adoption. Customer Service type teams where the work is continuous may lean toward Kanban style. Development teams may lean toward Scrum. If your organization is running a DEVSECOPS or SECDEVOPS type model, it may be more of a Scrum ban, which is a mixture of both. Remember, at the heart of Agile is the intent of being flexible and focusing on the most efficient means of getting a solution to market. Once you figure out the ideal approach for your team, standardizing the processes and procedures including a model with continuous improvement techniques built in helps to drive additional efficiencies.

More mature Agile teams focus on assessing industry trends by reviewing annual reports such as the State of Agile Report (Digital AI, 2024), which is published by digitalai. Program resources can assess process effectiveness and determine a strategy to improve current processes to proactively address trending concerns by reviewing the results of global surveys such as this one. This information can be discussed with scrum masters to disseminate to their teams annually. Maintaining a process that periodically reflects on the program to realign internal processes with the needs of the program is a good continuous improvement technique.

9 BUSINESS INTELLIGENCE AND ANALYTICS PROJECTS

The objective of the Business Intelligence and Analytics projects section is to focus on optimizing high performance teams on analytical projects and programs.

Primary Goals:

- Team Member Profiles
- Setting Customer Expectations
- Improving the Business and Technical Team Interaction

9.1 Business Intelligence and Data Analytical Solutions

Managing a high-performing team of individuals focused on business intelligence and data analytical solutions can be one of the most frustrating and satisfying experiences. The frustrating part is understanding the power of how the data can quickly achieve an organization's goal and limitations the team or others in the organization has based on a number of factors. Factors include being unaware of the team's capabilities, internal politics, and other strategic objectives only known by a few. Nothing, however, is more satisfying than leading one of these teams to success. Providing guidance to team members to build solutions and hearing the success from customers never gets old. This section dives a bit deeper into some of these topics. Many of the lessons learned here can be translated into tools for managing other types of teams.

9.1.1 Team Member Profiles

Information management and organization begins with understanding the important roles that lead to high performance. In most organizations, these roles have different management chains. The successful leadership team is one who understands that the mission is best achieved by understanding the three key roles and how to align each team's goal to better support the mission. For these teams, the three roles are that of the Business Process Owner, the Source Data Steward, and the Business Intelligence application owner. A Practical Guide for Implementing an Enterprise Information Management Program explored the importance of governance and building coalition to drive change by organizing teams to achieve success. Management of the enterprise information assets is core to any data related team, which includes data analytics and business intelligence programs.

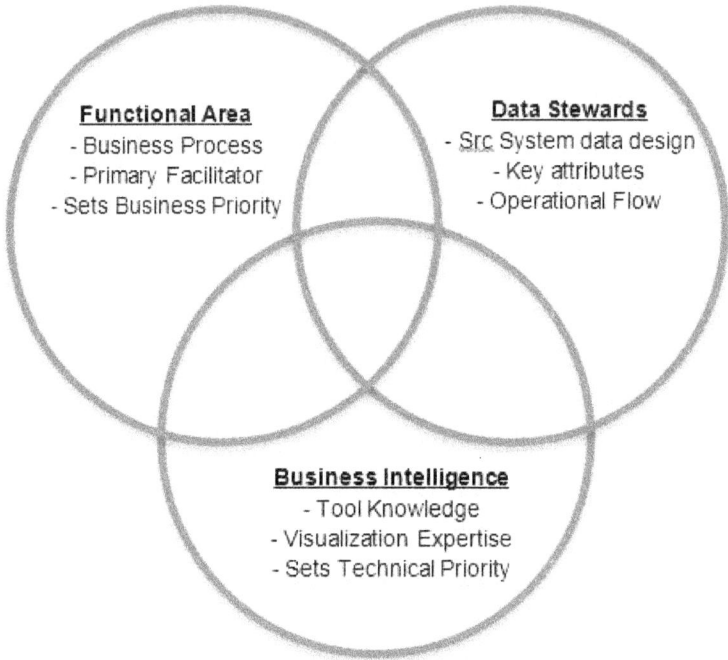

Functional Area
- Business Process
- Primary Facilitator
- Sets Business Priority

Data Stewards
- Src System data design
- Key attributes
- Operational Flow

Business Intelligence
- Tool Knowledge
- Visualization Expertise
- Sets Technical Priority

Figure 4: Business Intelligence Trifecta

Business Process Owners: The Business Process owners should always drive and be the primary focus of mission focused solution development. This person or team should be in the best position to understand the goal of the mission, the underlying business process, and how the business process aligns with achieving the objective.

Source Data Stewards: Good Source Data Stewards understand their systems thoroughly. In most teams that manage operational tactical applications, these individuals are aware of how the data is entered, stored and processed technically. Individuals on this team understand the fields and reference values.

Business Intelligence Team: This team is primarily responsible for building out solutions that provide the right data and insights to the right person at the right time given technical constraints. This team understands the tools available to turn the data into insight and help management make decisions that support their business processes.

More mature organizations have an Enterprise Information Management Program with strong governance that allows these three teams to flourish. Managed by a governance champion, the program liaisons across teams ensure that the business processes are documented, clear, and communicates continuous improvement. Processes include the approval and management of fields and reference values. This may include driving clear standards in how the data is stored, standardized naming convention, and code values with clear descriptions for reference data.

9.1.1.1 Understanding the Order of Operations, Key Fields, Dimensions/Valid Values, and How the Data is Maintained

Source Data Stewards are primarily responsible for understanding the order of operations. With this understanding, their focus is on implementing solutions that quickly implement quality data into the system. Understanding how data can be used effectively for decision-making helps to determine the key fields and those fields that are most pertinent for decision-making. With this understanding, they can focus on maximizing the quality of data provided for decision-making. Most systems where data enters the data life cycle have fields that are wide-open text fields. Modifying a system that minimizes the risk of bad data entering the data ecosystem should be a high priority.

9.1.1.2 Data Governance - How is the Data Maintained?

Determining how data is maintained should be done at the onset of any new solution. It is never too late to put in a good governance structure that allows the team to meet the mission of the organization. Governance should not get in the way of achieving an organization's goal. It should be lean enough to quickly assess changes that allow the organization to better meet its objectives. Understanding the Business Owners of the data is extremely important. Who determines what the valid values are for reference fields? What is the process for changing these values? What is the process for implementing new fields? What is the communication plan for such changes? If the answers to these questions are not clear across the organization, it is likely there is some room to grow in this area. In the worst-case scenarios, the new code must be rolled back due to reports being affected by a source field. In this scenario, stakeholders may not take into consideration the dependency of important reports that rely on the field, changed by the source system with the business owner's knowledge. The Business Intelligence Team should understand or have tools that provide data lineage from the analytics solution back to the source system. Many of the tools used to transform data provide reports and search capability for this reason.

9.1.1.3 Data Quality - Entering Standardized Values vs Free Form Text

Standardizing input helps to drive higher data quality standards that can be used for better decision making. Limiting the number of free text field input helps with this goal. If the business end user standardizes what they place in a free text field, it is often a good place to review and either create a field with new standardized entry or change the free text to standardize with the decision being dependent on the use case. Free text offers ways for a user to provide candid input that a system or process owner may not have considered. This could lead to enhancements in both the supporting business process and the underlying data, which may not be captured in a standardized field. Reference fields with standardized input do not usually allow flexibility for values not defined in the available list. One solution that some systems use is the 'Other' option where the standard value of other is provided leading to a free text field for entry. Teams with a continuous learning mindset will periodically review the order of operations, fields, supporting values and work with the Business Process Owners on any feedback received from end users. Doing this improves the business process flows and supports the order of operations and supporting fields including those with reference data.

9.1.4 Data Analytics Team

Optimizing a data analytics team requires patience, and as with most projects, a complimentary mix of perspectives and experienced resources who are passionate about data. Every team is unique and understands how to use the strengths of each to enrich the areas of growth for others and is a soft skill that can be learned through the right balance of training on soft and hard skills.

9.1.4.1 Team Makeup (Product Architect, PM/Scrum Master, Business Analyst)

Data analytic teams consist of three key personnel on the management level. The first is the Business Analyst. The Business Analyst role is best given to the person who continuously seeks to understand the business process, interprets important transitions in the process, and interprets the business requirements for the solution in a way that the technical team can easily develop solutions. The Product Architect is responsible for the technical solutioning. This includes setting standards for the development teams, designing a system that is sustainable for the expected growth of the organization, and maintaining sufficient information for securing and maintaining the solution's infrastructure.

The Project Manager or Scrum Master on Agile programs is primarily responsible for facilitating the solution build. The person in this role should have strong facilitation, communication, and organizational skills. In this role, they typically manage access to the team's collaboration site, schedule and facilitate most of the meetings, and most importantly, minimize risks by removing impediments and enhancing communication across stakeholders. The better the management team, the more value you can create with the development staff whose interest and focus should be on developing the most efficient codebase for analytical solutions. Complementing this management structure is the set of resources used to develop the solution. The team size may be driven by the amount of business intelligence vendor products maintained by the organization. There is usually at least one expert in each business intelligence tool whose role is to interact with the vendor and determine the best features relevant to the business scenarios provided. This subject matter expert understands the significance of functional features when products are upgraded and helps to set direction by providing the technical complexity of requested solution components. Understanding the technical complexity of solutions helps to drive the implementation plan of the solution. See *Section 8.2.2 Backlog Grooming* for more information.

9.1.4.2 Understanding Data for Visualization - Metrics and Dimensions

Any member of the team interacting with the customer should have a basic understanding of the components utilized in creating good visualization and dashboards. If you are managing a Business Intelligence team, it is important that annual training includes refreshers on updated standards, processes, and capability that is important to use to build consistency in the data analytical solutions. Periodic assessment of standards and continuous improvement opportunities allow for consistent delivery to the customer. Consistent delivery leads to lowering the learning curve of understanding the solutions provided and completion of the mission.

9.1.4.2.1 Metrics

Metrics are the key facts used to measure performance. It may include the number of widgets that management would like to measure. This could be patients entering a hospital, cases a caseworker completes, automobiles built in a factory, new customers joining the company, financial information for the organization financial statements etc.

9.1.4.2.2 Dimensions

Dimensions are other attributes of the data that can be measured. Some of the more common dimensions are time, location, organization, type and status. Some of these may even be broken down into more detailed groups. Much of this depends on what the management team is trying to measure to effect change and compliance to service level agreements (SLAs).

9.1.4.2.3 Business Intelligence Target Customer (Supervisor vs Team member – queue)

Understanding the target customer and their role in the business process helps to identify which standard should be used for development. For example, are you building a solution for a supervisor? The focus of management is to continuously make things better for the team. The standard may include the number of items being built or addressed, how many items each team member is building over a specified time frame, or how successful is the team on meeting its goal? Where are the constraints in the business process? Most of the time it is reflective, backward-looking, to help determine changes that can be made to make things better, faster, with minimum constraints. Everything around the metrics is about measuring to determine improvement opportunities and success in maximizing the value the business process provides. Alternatively, if the target is a team member, the solution is most likely similar to a work queue which provides a list of tasks assigned to be accomplished. The objective is different.

9.1.4.3 Standardizing the Layout for the Dashboards

Once we understand the types of audience, the next step is to understand the dashboard layout that resonates best for the audience

216

type. As a group, do they prefer a light background or dark background? Do they print the items to review on paper or are they comfortable making decisions by electronic dashboards only? What is their focus? Are they interested in increasing the team's efficiency or are they measuring revenue? Building dashboards and data analytical solutions that complement decision makers' processes is a key to success on any of these programs. Aligning the information provided on the dashboard with the key performance indicators of the organization is a result driven approach to building continuous improvements and success with your team.

10 OPERATIONS AND MAINTENANCE

The objective of the Operations and Maintenance section is to provide tools to optimize a high-performance team responsible for maximizing system availability while reducing risks in matured systems after development is completed.

Primary Goals:

- Preparing for Support (Distribution Lists Creation/Periodic Review)
- Customer Support Monitoring
- License Management
- Vulnerability Monitoring and Remediating
- Infrastructure Maintenance Windows (Upgrades, Patches, Releases)

10.1 Preparing for Support

Managing an Operations and Maintenance team can be one of the most stressful roles in an organization. It is also an area where the management team can provide great value, creating a structured environment and controlling risk. The Operations and Maintenance team is a thankless job. Often, customers only are aware and acknowledge team members in times of high stress, when the system is not working according to plan, and when they need to meet a deadline. Growing these resources and optimizing a high-performance team in this area requires a slightly different approach than a development team. Operations and Management Team members are primarily focused on system stability which can be optimized by focusing and automating processes and procedures for communication, monitoring customer support, monitoring the response time, and configuring the customer support team to adequately support the customer base.

Each distribution list should have a purpose and there should be a periodic review of all distribution lists of no less than once a year. Distribution lists with the purpose and tips on what to look for should be documented in a centralized location. This easily accessible document can be leveraged for new resources and for creating good transition plans.

Anecdote: On one occasion, after successfully managing a large development team, I was asked to take over the Operations and Maintenance team for a significant program that was going up for rebid in one year. Many of the strong management resources on the team chose to leave due to alliances with a potential competitor who was a former leader on the program. As such, quickly forming a management team to address a stabilized steady unit was paramount. This team, not knowing my background, was skeptical about the new program manager promoting me to be the deputy for this team of over thirty individuals. I still remember the first week; after one of the primary managers resigned, asking me if I was sure I wanted to receive all the automated system generated emails since they received hundreds of emails a day. "Yes", I said. The experience taught me two things. First, be careful what you wish for. Second, your training has prepared you for managing most situations. Knowing that email frequency would be challenging, one of the first things I did was to automate the messages coming into my mailbox by distribution list. My second action was to work with this individual to understand the day-to-day cadence to better understand the importance of the various distribution lists. It allowed me, most importantly to establish a create a strong transition plan which was reused after recruiting another very talented replacement resource for the position. I let the new technical resource focus on the technical nuts and bolts part of the role, while I established good supporting procedures and techniques. Every member of this management team continued their professional growth over the year and each one of them were better professionals who went on to do great things in their respective new positions. They were all life-long learners committed to a mission, who allowed us to see a major cloud migrating transformation through to the end.

10.2 Customer Support Monitoring

Customer support monitoring is very important in understanding team effectiveness in developing solutions that meet the needs of the customer. If an organization has a help desk group, understanding how to tier the help desk resources is important. Scripts can be created in tiers. The first tier is the customer facing resources. Scripts can be created to follow simple instructions that address most of the novice users of the solution. Individuals in tier one must be able to quickly assess the customer's needs, remediate, and when possible, communicate what the team is doing and provide an estimate on when they will hear back from someone. The technical depth of these resources does not have to be extensive. What is more important is their ability to communicate well with both a technical and non-technical customer. Tier two customers do require some technical depth. One approach is to use tier two to grow a resource from tier one, and similarly for tier two resources and having a growth path for the individual to be part of the development team. Tier three should be primarily strong developers who are independently knowledgeable and can solve tough customer issues because they know the code well.

One technique is to perform an assessment of the customer support tickets to look for patterns and opportunities for improvement. Understanding why the users are submitting tickets allows the program office to provide better information to the correct stakeholder. Customer training materials may be improved by including a frequently asked questions section with issues that can be easily resolved by educating the end user. Training guides and support scripts can be updated and provided to the customer support resources. Improvements in internal processes can be tied directly to the team's performance. Given sufficient transition time, the management resources of the operations and maintenance group can take advantage of having a temporary surge of the transition resource to review and fix key processes that will help the team in the end.

10.3 License Management

License management is necessary on large complex programs with multiple applications. Usually, the application technical lead is responsible for ensuring the necessary software is requested, received, and periodically updated based on the application and organization requirements. This one area, if consolidated and mature, can decrease the number of incidents and minimize product outages. Commercial off-the-shelf applications usually have a fixed update schedule and a process for urgent fixes or patches. Understanding these cycles allows the operation and maintenance team to place in a calendar for visibility and build automated notification routines to manage during regular maintenance upgrades. If all the licenses and patches are in one place, the team can quickly confirm the upcoming maintenance work and ensure they are proactive in assigning and covering the work in the periodic maintenance window.

10.4 Vulnerability Monitoring and Remediating

Monitoring and remediating vulnerabilities have received much visibility over the years. Years ago, organizations assumed their purchased solutions were secure. Vulnerability scans were used reactively and only identify infected systems. Today, there is an entire industry dedicated to securing the organization's infrastructure. The industry then moved to DevSecOps to highlight security, being an integral part of development before moving to operations. Now security is an integral part of the development process and is considered before development begins. There are several techniques that can be utilized to better secure a technical environment and it starts with hiring or training the right set of resources. For example, the federal government requires roles to have certifications such as the Certified Information Systems Security Professional (CISSP), as a base requirement for resources on these teams. As a result, the private sector follows these guidelines as a minimum requirement for the specified support staff if they want to do business with the government. Management in this area of operations and maintenance usually requires at least a CISSP and over five years of experience remediating security vulnerabilities. More mature teams have dashboards specifically looking at each division, the level of outstanding security vulnerability findings, and track how long it is taking teams to remediate these vulnerability findings once identified. Implementing policy around remediation time, depending on the level of criticality, is essential to make efficient use of time. For example, no new solution should be allowed if there are known critical or high risks. Any critical risks found in production should be remediated at the next production maintenance window unless management communicates an immediate fix is required. These usually affect the availability of the solution to the customers who are providing mission critical services, so management needs to weigh the chances and implications of the risk against the criticality of the mission being served. More mature organizations have a window of at least a couple hours a week to perform regular system maintenance that now includes security monitoring and hardening.

10.5 Infrastructure Maintenance Windows

The infrastructure environments are one area of massive improvement over the years. Today, there is little need for system downtime if an organization has the money and the resources to set up a technical infrastructure solution correctly. With that said, there are many organizations who do not require or cannot afford a twenty-four by seven-uptime solution. Either way, programs that have both operations and maintenance teams should always reserve the lowest utilized period to perform periodic maintenance and place it on a known schedule. It could be used for batch processing, vulnerability remediation, application patching, and upgrades, etc. Ideally, you would create a schedule that allows you to quickly implement emergency fixes. This schedule varies by organization and it depends on your global reach. A smaller organization doing business only on the east coast will have much more flexibility than an organization that works across the country. Likewise, an organization with national reach will have more flexibility than those who service global customers.

The Operation & Maintenance team is the one team where you can control the destiny of any large technical solution implementation. Preparing the management team to address the large quantity of automated messages is the first step on optimizing continuous improvement efficiencies into the process as it allows the management team to quickly monitor and triage a large number of notifications. Monitoring and periodically assessing customer support tickets allows the management team to build a structure that effectively meets the needs of the customer base with the right mix of team members. Preparing processes and procedures to proactively manage a license monitor and remediate vulnerabilities while setting up pre-defined windows for recurring maintenance helps in optimizing this high-performing team.

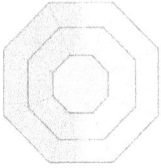

SUMMARY

Implementing technical solutions is difficult and not insurmountable. Early in my career, I was fortunate to work for the US Virgin Islands Territorial Court System. This court system sponsored a pre-trial initiative for youth with the motto: nothing is so complicated that it cannot be accomplished by hard work. The motto is fitting for an instrumental steel band who has many individuals playing different types of instruments to create music for fans to enjoy. Technical project management is similar in concept. It is difficult, with many complexities that, once put together correctly, can create value for organizations and allow them to be successful in their mission.

The first section of the book addressed the infrastructure required to be successful in the organization's journey. Starting with the end in mind, the organization creates an environment that supports continuous learning with clear standards and procedures that will be scalable to grow and expand with the program. The beginning of the journey utilized tools like assessment surveys to better understand the organization's maturity across six focus areas. Choose a tool that is best suited to your organization and we have provided recommendations, such as the **PMO Organization Assessment,** in the appendix for your reference. The importance of a collaboration site and central library, a place that houses a repository of the team's reference and working documents, becomes incrementally more critical as the team continues to expand. All documents and the organization of the central library should be driven by industry standards. Setting expectations of the team with an interaction model that allows for professional growth in a safe and trustworthy environment, and measuring success through predetermined key performance indicators allows the team to have a gauge of progress throughout the program life cycle. Finally, it is important to align personal and professional goals and saying, "Thank you", to show appreciation to the team and the individual team members who are committed to the program.

After the initial program infrastructure was in place, the next part of the journey was focusing on Business Development. The objective in this section was to optimize the process of generating leads and turning those leads into value-added revenue streams. The focus was on opportunity building and proposal team management. Areas in this section include management of leads and prospects, optimizing the team size for capturing new work, creating a standard approach to responding to new opportunities, and measuring performance of your business development team.

Next, the book explored techniques for optimizing high-performing teams quickly. Recruitment Management was the focus as it delved into candidate sourcing and recruitment team management. The correct team makeup for effective recruiting, aligning the schedule of key stakeholders, creating standard operating procedures for the team and measuring recruitment efforts are all important concepts. The importance of hiring a diverse team similar in their commitment to life-long learning and success in their craft cannot be overstated.

After creating a framework for the program infrastructure and recruiting, the focus turned to how best to integrate the team through communication, requirements, and capturing continuous improvement opportunities. The goals included managing stakeholder expectations, developing a standards-based code set, establishing procedures to minimize technical depth, using risk and quality to mitigate project risks, and measuring outcomes.

With the framework for integration in place, we focused on Solution Development. In this section, we explored the development environment and techniques for developers to quickly access information they need. We discussed the importance of development standards, developing with a security mindset, managing risks across the development team, using standard development frameworks such as Agile, and measuring solution development for continuous development opportunities.

Once we get our technical solutioning framework in place, we cannot forget long-term caring and feeding to optimize the value our technical solutions provide. We discussed both production and customer support techniques. We touched on the tools that help us to prepare for support of rapidly growing and changing systems, customer support monitoring, license management, vulnerability monitoring and remediation, and infrastructure maintenance windows.

A common theme across each of these areas is how best the organization can use a standards-based approach and tools to make process improvements, and to create a culture that focuses on specific improvements to maximize our strategic goals. This reference guide was organized to provide techniques used throughout a technical program management life cycle, from techniques used in program office setup through standardized tools and techniques for maintaining and managing technical solutions. It is by no means detailed to cover every scenario and, rather, is a starting point for you to determine the next step in your journey. Each of us is in a particular place in our journey, whether starting up a program, or seeking to make a mature program even more effective; this reference manual can be used to focus on the specific areas as part of that journey. The assessments can also help you identify where to focus for your strategic implementation plan.

One note on disruptors. There is another saying in project management that the only known truth is that the schedule you create at the beginning is not the same you end up with at the close of the project. This is primarily because of the unknowns or disruptors. It may be simple things like a shift in business based on customer demands, changes to development life cycle like the new Agile 2.0, new technology waves like moving to the cloud, or Artificial Intelligence and Machine Learning, or even unexpected events like 9/11 that caused a new shift to focus on Disaster Recovery and Business Continuity Planning. One point to drive home is that there will always be disruptors. Understanding this, optimizing high performance teams requires providing the team with the skills they need to support an ever-changing landscape in technology. There will always be inside and outside factors that affect the priorities of the business. Having a prioritized list of initiatives that is periodically reviewed allows the program to keep up to the speed of business, one of the reasons Agile has become so successful.

Anecdote: As I continue my professional journey in technical project management, my hope is that my readers take away a few things that help them in both their professional and life's journey. I've learned that success is not driven alone by the completion of individual projects or missions, but in the journey itself. Cherishing the people you meet along the way and helping each other to grow and be productive members of society. I encourage you to continue your learning, helping others to grow, and together placing our piece of the puzzle in this journey called 'life'. Be intentional about your journey. Surround yourself with a talented team who is mission focused, diversified and committed to a lifetime of learning, sharing knowledge, and continuous improvement. Leverage the vast content available to select the best tools to help you and your team become successful. Good luck on your journey and I hope to meet you at some point along the way.

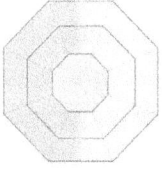

APPENDIX

The Appendix provides additional information that supports the prior sections. The assessments and templates sections provide a link to more in-depth opportunities to advance your knowledge within your specific organization.

- Acronyms
- Assessments (with Link)
- Templates (with Link)
- Index
- About the Author

Appendix 1: Acronyms

Below is a list of commonly used acronyms that is a good reference for high performing team members.

Acronym	Terminology
508	The Section 508 Standards, which are part of the Federal Acquisition Regulation, ensure access for people with physical, sensory, or cognitive disabilities.
6 Sigma	Six Sigma Certification
a-IPC	Associate - Infection Prevention and Control
AB	Accreditation Body
ACA	Affordable Care Act
ACCA	Association of Chartered Certified Accountants
ACD	Applied Cybersecurity Division
ACEP	American College of Emergency Physicians
ACMP	Association of Change Management Professionals
ACO	Accountable Care Organizations
ACQ	Acquisition
AHA	American Hospital Association
AHA-CC	American Hospital Association - Certification Center
AHRQ	Agency for Healthcare Research and Quality
AIA	American Institute of Architects
AICPA	American Institute of Certified Public Accountants
AIMA	Alternative Investment Management Association
AMA	American Medical Association
ANSI	American National Standards Institute

Acronym	Terminology
AoA	Analysis of Alternatives
AoNL	American Organization for Nursing Leadership
APEC	Asia-Pacific Economic Cooperation
APHA	American Public Health Association
APMP	Association of Proposal Management Professionals
APMP BOK	Association of Proposal Management Professionals Book of Knowledge
APQP	Advanced Product Quality Planning
ARF	Asset Reporting Format
ASC	Accredited Standards Committee
ASCII	American Standard Code for Information Interchange
ASCLD/LAB	The American Society of Crime Laboratory Directors/Laboratory Accreditation Board
ASVS	OWASP Application Security Verification Standard
AUP	Acceptable Use Policy
BA	Business Analyst
BCP	business continuity plan
BI	Business Intelligence
BPaaS	Business Process as a Service
BRM	Business Reference Model
BRMI	BRM Institute
BRMP	Business Relationship Management Professional
CAB	Change Advisory Board
CAE	Chief Audit Executive
CAMS	Culture, Automation, Measurement, and Sharing

Acronym	Terminology
CAP	Corrective Action Plan
CAP APMP®	APMP Capture Practitioner Certification
CAPM®	Certified Associate in Project Management
CAS	Cost Accounting Standards
CBA	Cost-benefit analysis
CBIC	Certification Board of Infection Control and Epidemiology, Inc
CBRM	Certified Business Relationship Manager
CBT	Computer-based training
CC	Common Criteria
CCB	Change (or Configuration) Control Board
CCM™	Certified Capture Manager
CCMP™	Certified Change Management Professional
CCTA	Central Computer and Telecommunications Agency
CDC	Centers for Disease Control and Prevention
CDI	Center for Data Innovation
CDMP®	Certified Data Management Professional
CDO	Chief Data Officer
CDR	Critical Design Review
CENP	Certified in Executive Nursing Practice
CEO	Chief Executive Officer
CER	Crossover Error Rate
CERT-UK	Computer Emergency Response Team
CF APMP®	APMP Foundation-Level Certification
CFO	Chief Financial Officer

Acronym	Terminology
CFR	Code of Federal Regulations
CFS	Cybersecurity Framework
CGFNS	CGFNS International
CHC	Certified Healthcare Constructor
CHCIO	CHIME Certified Healthcare CIO program
CHESP	Certified Healthcare Environmental Services Professional
CHFM	Certified Healthcare Facility Manager
CHHR	Certified Healthcare in Human Resources
CHIME	College of Healthcare Information Management Executives
CIA	Central Intelligence Agency
CIC	Certified in Infection Control
CIHI	Canadian Institute for Health Information
CIO	Chief Information Officer
CIPM	Certified Information Privacy Manager
CIPP	Certified Information Privacy Professional
CIPT	Certified Information Privacy Technologist
CIR	Committed Information Rate
CIS	Center for Internet Security
CISA	Cybersecurity & Infrastructure Security Agency
CISM	Certified Information Security Manager
CISO	Chief Information Security Officer
CiSP	Cyber Security Information Sharing Partnership (UK)
CISSP	Certified Information Systems Security Professional
CJIS	Criminal Justice Information Services

Acronym	Terminology
CLO	Chief Legal Officer
CM	Configuration management
CMMC	Cybersecurity Maturity Model Certification
CMMI	Capability Maturity Model Integration
CMMI-A	Certified CMMI® Associate
CMMI-P	Certified CMMI® Professional
CMRP	Certified Materials & Resources Professional
CMS	Centers for Medicare & Medicaid Services
CNI	Critical Network Infrastructure
CNML	Certified Nurse Manager and Leader
CNSS	Committee on National Security Systems
CNSSD	Committee on National Security Systems Directives
CNSSI	Committee on National Security Systems Instruction
COBIT	Control Objectives for Information and Related Technologies
CONOPS	Concept of Operations
COO	Chief Operating Officer
COOP	Continuity of Operations Plan
COSO	Commission of Sponsoring Organizations of the Treadway Commission
COTS	Commercial off-the-shelf
CP APMP®	APMP PRACTITIONER-LEVEL CERTIFICATION
CPA	Certified Public Accountant
CPFP	Cost Plus Fixed Price
CPHRM	Certified Professional in Healthcare Risk Management

Acronym	Terminology
CPO	Chief Privacy Officer
CPP APMP®	APMP-PROFESSIONAL™ LEVEL
CR	Change Request
CRAMM	CCTA Risk Analysis and Management Method
CRM	Customer Relationship Management
CRO	Chief Risk Officer
CSA	Cloud Security Alliance
CSA	Cloud services agreement
CSAIC	Cyber Security and Information Systems Information Analysis Center
CSD	Computer Security Division
CSE	Communication Security Establishment
CSIRT	Computer Security Incident Response Team
CSM	Certified Scrum Master
CSO	Chief Security Officer
CSP	Content Security Policy
CSRC	COMPUTER SECURITY RESOURCE CENTER
CSV	Comma-separated values
CTCPEC	Canadian Trusted Computer Product Evaluation Criteria
CTO	Chief Technology Officer
CUSP	Comprehensive Unit-based Safety Program
CVE	Common Vulnerabilities and Exposures
DA	Data Architect
DAC	Disciplined Agile Coach

Acronym	Terminology
DAM	Digital Asset Management
DAM	Database Activity Monitoring
DAMA	DAMA International
DASM	Disciplined Agile Scrum Master
DASSM	Disciplined Agile Senior Scrum Master
DAVSC	Disciplined Agile Value Stream Consultant
DBA	Database Administrator
DBaaS	Database as a Service
DBMS	Database Management System
DC3	DoD Cyber Crime Center
DCE	Distributed Computing Environment
DCMA	Defense Contract Management Agency
DEV	Development
DEVSECOPS	Development, Security, and Operations
DFARS	Defense Federal Acquisition Regulation Supplement
DHS	Department of Homeland Security
DISA	Defense Information Systems Agency
DM	Data Migration
DMM	Data Maturity Model
DMP	Data Migration Project
DoD	Department of Defense
DoD SRG	Department of Defense Cloud Computing Security Requirements Guide
DOJ	Department of Justice

Acronym	Terminology
DORA	DevOps Research and Assessment
DPA	Data Protection Act
DRP	Disaster Recovery Plan
DUA	Data Use Agreement
DAST	Dynamic Application Security Testing
EAL	Evaluation Assurance Levels
eCFR	Electronic Code of Federal Regulations
ECOSOC	Economic and Social Council
ECP	Engineering Change Proposal
EGIT	Enterprise Governance of Information and Technology
EHR	Electronic Health Record
EIM	Enterprise Information Management
EIS	Enterprise Information System
EOL	End-of-Life
EPA	Environmental Protection Agency
ERM	Enterprise Risk Management
EVM	Earned Value Management
FAR	Federal Acquisition Regulation
FAR	False Acceptance Rate
FBI	Federal Bureau of Investigation
FCC	Federal Communications Commission
FDA	Food and Drug Administration
FDCC	Federal Desktop Core Configuration
FEAF	Federal Enterprise Architecture Framework

Acronym	Terminology
FedRAMP	Federal Risk and Authorization Management Program
FERPA	Family Educational Rights and Privacy Act
FFIEC	Federal Financial Institutions Examination Council
FFP	Firm Fixed Price
FHIR	Fast Healthcare Interoperability Resource
FIPS	Federal Information Processing Standards
FIPS	Federal Information Security Modernization Act
FISCAM	Federal Information System Controls Audit Manual
FISMA	Federal Information Security Modernization Act
FISMA	Federal Information Security Modernization Act
FMEA	Failure Modes and Effects Analysis
FMG	FHIR Management Group
FOIA	Freedom of Information Act
FRR	False Rejection Rate
FSO	Facility Security Officer
FTC	Federal Trade Commission
FTE	Full Time Equivalent
GAAP	Generally Accepted Accounting Principles
GAO	Government Accountability Office
GAPP	Generally Accepted Privacy Principles
GDPR	General Data Protection Regulation
GFE	Government Furnished Equipment
GLBA	Gramm-Leach-Bliley Act
GovWin	Government Win Intelligence Platform

Acronym	Terminology
GRC	Governance, Risk Management, and Compliance
GxP	Good X Practice
HCPCS	Healthcare Common Procedure Coding System
HHS	Health and Human Services
HICS	Hospital Incident Command System
HIPAA	Health Insurance Portability and Accountability Act
HITECH	Health Information Technology for Economic and Clinical Health
HITRUST	Regulatory Compliance & Risk Management Framework
HL7	Health Level Seven
HL7 CDA	HL7 CDA Specialist
HL7 FHIR Proficient	HL7 FHIR Proficient
HL7 V2	HL7 V2 Control Specialist
HL7 V3 RIM	HL7 V2 RIM Specialist
HMO	Health Maintenance Organization
HR	Human Resources
I3A	International Imaging Industry Association
IAPP	International Association of Privacy Professionals
IASSC	The International Association for Six Sigma Certification
IAST	Interactive Application Security Testing
ICD	Interface Control Document
ICD (version #)	International Classification of Diseases, Version #
ICO	Information Commissioner's Office
ICREA	International Computer Room Experts Association

Acronym	Terminology
IDEA	International Data Encryption Algorithm
IDS	Intrusion Detection System
IPS	Intrusion Prevention System
IEC	International Electrotechnical Commission
IEEE	Institute of Electrical and Electronics Engineers
IETF	Internet Engineering Task Force
IHE	International Health Exchange
ILM	Information Lifecycle Management
IMDRF	International Medical Device Regulators Forum
IMIA	International Medical Informatics Association
IMS	Integrated Master Schedule
INCOSE	International Council of Systems Engineering
INTERPOL	International Criminal Police Organization
IPMA	International Project Management Association
IRS	Internal Revenue Service
IRS 1075	Tax Information Security Guidelines
ISA	International Society of Automation
ISA/IEC 62443	Cybersecurity Certificate Program
ISACA	Information Systems Audit and Control Association
ISACs	Information Sharing and Analysis Centers
ISAE	International Standard on Assurance Engagements
ISAGCA	ISA Global Cybersecurity Alliance
ISAOs	Information Sharing and Analysis Organizations
ISCM	Information Security Continuous Monitoring

Acronym	Terminology
ISO	International Organization for Standardization
ISO27001	International Standard - Information Security Management
ISO27002	International Standard - Security Standards and Controls
ISO27017	International Standard - Security Techniques - Code of Practice for the Cloud
ISO27018	International Standard - PII in public clouds
ISO31000:2018	International Standard - Risk Management
ISO9001	International Standard - Quality Management
ISSE	Information Systems Security Engineer
ISSEP	Information System Security Engineering Professional
ISSM	Information System Security Manager
ISSO	Information System Security Officer
ISSP	Information System Security Program
IT	Information Technology
ITA	Information Technology Agreement
ITAM	Information Technology Asset Management
ITAR	International Traffic in Arms Regulations
ITIF	Information Technology and Innovation Foundation
ITIL®	Information Technology Infrastructure Library
ITIL® - F	ITIL® Foundation Certification
ITIL® - M	ITIL® Master Certification
ITIL® MP	ITIL® 4 Managing Professional
ITIL® SL	ITIL® 4 Strategic Leader
ITSEC	Information Technology Security Evaluation Criteria

Acronym	Terminology
ITSM	Information Technology Service Management
ITU	International Telecommunication Union
ITU-T	International Telecommunication Union - Telecommunication Standardization Sector
JPEG	Joint Photographic Experts Group
JSA	Japanese Standards Group
JTAG	Joint Test Action Group
KPI	Key Performance Indicator
KRI	Key Risk Indicator
LM	Lan Manager
LMS	Learning Management System
LOE	Level of Effort
MAC	Mandatory Access Control
MAN	Metropolitan Area Network
MDAAP	Medical Device Single Audit Program
MDM	Master Data Management
MFA	Multi-Factor Authentication
MITA 3.0	Medicaid Information Technology Architecture
MLS	Multilevel Security
MOA	Memorandum of Agreement
MOU	Memorandum of Understanding
MSA	Master Service Agreement
MSP	Managed Service Provider
MSSP	Managed Security Service Provider

Acronym	Terminology
MTBF	Mean Time Between Failures
MTD	Maximum Tolerable Downtime
MTPOD	Maximum Tolerable Period of Disruption
MTTF	Mean Time to Failure
MTTR	Mean Time to Repair
MTTR	Mean Time to Recovery
MVP	Minimum Viable Product
NARA	National Archives and Records Administration
NCC	National Cybersecurity Center
NCSC	The National Cyber Security Centre (UK)
NDA	Non-Disclosure Agreement
NFS	Network File System
NHS	National Health Service
NHSN	National Healthcare Safety Network
NIACAP	National Information Assurance Certification and Accreditation Process
NIAP	National Information Assurance Partnership
NICE	National Initiative for Cybersecurity Education
NIEM	National Information Exchange Model
NIST	National Institute of Standards and Technology
NPV	Net Present Value
NSA	National Security Agency
NSC	National Security Council
NSS	Network Security Services

Acronym	Terminology
NSTISSAM	National Security Telecommunications and Information Systems Security Advisory Memorandum
NSTISSI	National Security Telecommunications and Information Systems Security Instruction
NSTISSP	National Security Telecommunications and Information Systems Security Policy
NVD	National Vulnerability Database
OAS	Open API Specification
OCR	Office for Civil Rights
OCTAVE	Operationally Critical Threat, Asset, and Vulnerability Evaluation
OEM	Original Equipment Manufacturer
OEP	Occupant Emergency Plan
OIG	Office of the Inspector General
OLA	Operational-Level Agreement
OMB	Office of Management and Budget
ONC	Office of the National Coordinator for Health Information Technology
OOO	Out of Office
OPC	OWASP Top 10 Proactive Controls Project
OPEX	Operational Expenditure
ORCL	Oracle
ORR	Operational Readiness Review
OSD	Office of the Secretary of Defense
OWASP	Open Web Application Security Project
P-ATO	Provisional Authorization to Operate

Acronym	Terminology
PACS	Philanthropy and Civil Society - Standford
PBAC	Policy-based Access Control
PCI-DSS	Payment Card Industry Data Security Standard
PCI-DSSC	Payment Card Industry Data Security Standard Council
PCP	Primary Care Provider
PDDQ	Patient Demographic Data Quality Framework
PDF	Portable Document Format
PERT	Program Evaluation and Review Technique
PFDD	Patient Focused Drug Development
PfMP®	Portfolio Management Professional
PgMP®	Program Management Professional
PHI	Protected Health Information
PHR	Personal Health Record
PIA	Privacy Impact Assessment
PII	Personally Identifiable Information
PIN	Personal Identification Number
PIPEDA	Personal Information Protection and Electronic Documents Act
PM	Project Manager; Program Manager
PMBOK	Project Management Body of Knowledge
PMI	Project Management Institute
PMI-ACP®	PMI Agile Certified Practitioner
PMI-PBA®	PMI Professional in Business Analysis
PMI-RMP®	PMI Risk Management Professional
PMI-SP®	PMI Scheduling Professional

Acronym	Terminology
PMO	Program Management Office
PMP®	Project Management Professional
PNG	Portable Network Graphics
POA&M	Plan of Action and Milestones
POC	Proof of Concept
POC	Point of Contact
PoLP	Principle of Least Privilege
PoT	Proof of Technology
PQQ	Pre-Qualification Questionnaire
QA	Quality Assurance
QoP	Quality of Protection
QoS	Quality of Service
QSA	Qualified Security Assessor
R&D	Research and Development
RACI	Responsible, Accountable, Consulted, and Informed
RAD	Rapid Application Development
RASIC	Responsible, Accountable, Supportive, Informed, Consulted
RASP	Runtime Application Self-Protection
RBAC	Role-Based Access Control
RBIA	Risk-Based Internal Audit
RCA	Root Cause Analysis
RDS	Relational Database Services
REST	Representational State Transfer
RFC	Request for Commitment

Acronym	Terminology
RFI	Request for Information
RFP	Request for Proposal
RFQ	Request for Quote
ROE	Return on Equity
ROI	Return on Investment
ROSI	Return on Security Investment
RPO	Recovery Point Objective
RTM	Requirements Traceability Matrix
RTO	Recovery Time Objective
S/MIME	Secure/Multipurpose Internet Mail Extensions
SA	System Architect
SaaS	Software as a Service
SAFECode	Software Assurance Forum for Excellence in Code
SAISO	Senior Agency Information Security Officer
SAL	Security Assurance Level
SAM	System for Award Management
SAML	Security Assertion Markup Language
SAMM	Software Assurance Maturity Model
SAQ	Self-Assessment Questionnaire
SAR	Security Assessment Report
SAS	Statement on Auditing Standards
SAST	Static Application Security Testing
SBA	Small Business Administration
SCA	Software Composition Analysis

Acronym	Terminology
SCM	Software Configuration Management
SCM	Source Code Management
SCRM	Supply Chain Risk Management
SDL	Security Development Lifecycle
SDLC	Software Development Life Cycle
SDLC	Systems Development Life Cycle
SEC	Securities and Exchange Commission
SECDEVOPS	Security, Development and Operations
SEC Rule 17-a-4(f)	Data Retention
SEM	Security Event Management
SEMP	System Engineering Management Plan
SEP	Systems Engineering Plan
SETA	Security Education, Training, and Awareness
SIEM	Security Information and Event Management
SIL	Safety Integrity Level
SIM	Security Information Management
SLA	Service-Level Agreement
SLR	Service Level Requirement
SMART	Specific, Measurable, Achievable, Relevant, and Time-Bound
SME	Subject Matter Expert
SOA	Service Oriented Architecture
SOC	System and Organization Controls
SOHO	Small Office/Home Office

Acronym	Terminology
SOO	Statement of Objectives
SOP	Standard Operating Procedure
SoR	System of Record
SORN	System of Records Notice
SOX	Sarbanes-Oxley
SPF	Sender Policy Framework
SPI	Schedule Performance Index
SPI	Security Parameter Index
SQL	Structured Query Language
SQLi	SQL Injection
SRR	System Requirements Review
SRS	Software Requirements Specification
SRTM	Security Requirements Traceability Matrix
SSAA	System Security Authorization Agreement
SSAE	Statements on Standards for Attestation Engagements
SSDF	Secure Software Development Framework
SSN	Source System Notice
SSO	Single Sign On
SSP	System Security Plan
STIG	Security Technical Implementation Guide
SSIR	Standford Social Innovation Review
SVC	Service
SWOT	Strengths, Weaknesses, Opportunities, and Threats
T&M	Time and Material

Acronym	Terminology
TaaS	Testing as a Service
TAG	Technical Advisory Group
TARA	Threat Agent Risk Assessment
TARA	Threat Assessment and Remediation Analysis
TCO	Total Cost of Ownership
TDD	Test-Driven Development
TEMP	Test and Evaluation Master Plan
TIFF	Tagged Image File Format
TIP	Technology Innovation Program
TOGAF	The Open Group Architecture Framework
TPA	Third-Party Administrator
TPM	Technical Performance Measures
TRR	Test Readiness Review
TST	Testing
TTP	Trusted Third Party
UAT	User Acceptance Testing
UI	User Interface
UMLS	Unified Medical Language System
UNI	Ente Nazionale Italiano di Unificazione
UNIT	Instituto Uruguayo De Normas Technicas
UPC	Universal Product Code
URI	Uniform Resource Identifier
US	United States
US-CERT	United States Computer Emergency Readiness Team

Acronym	Terminology
USAB	United States Access Board
USAGOV	United States of America Government
USGCB	United States Government Configuration Baseline
USPS	United States Postal Service
UTM	Unified Threat Management
VM	Virtual Machine
VoIP	Voice Over Internet Protocol
VP	Vice President
VPAT™	Voluntary Product Accessibility Template
VPC	Virtual Private Cloud
WAF	Web Application Firewall
WAP	Wireless Access Point
WBS	Work Breakdown Structure
WHO	World Health Organization
WiMAX	Worldwide Interoperability for Microwave Access
WIPO	World Intellectual Property Organization
WORM	Write Once, Read Many
WPS	Wi-Fi Protected Setup
WRT	Work Recovery Time
XACML	eXtensible Access Control Markup Language
XAML	Transaction Authority Markup Language
XML	Extensible Markup Language
YAML	YAML Ain't Markup Language
ZTA	Zero Trust Architecture

Appendix 2: PMO Maturity Assessment

A maturity assessment has been created to allow you to determine which areas of a project management office may be most applicable for you. The assessment can be accessed by going to the following site: PMO Maturity Assessment. We encourage you to take the assessment using code: **PMRAS25** so that we can provide a summary of the most useful templates for our readership.

Program Management Rapid Assessment Survey (PMRAS) provides a quick smart assessment for identification and planning of recommended short-term initiatives. The Program Management Detail Assessment Survey (PMDAS) is a more thorough organizational assessment to help with short-, mid- and long-term planning in an organization.

Appendix 3: My Digital Libraries - Addendum

There are many great sites to obtain templates that work best for you. I highly recommend starting with templates/standards produced by your organization. One measurement of an organization's maturity is the level of rigor in the procedures and templates that are institutionalized. Another great source for templates includes organizations such as PMI, ITIL, ISO, etc. Leading practice templates can be recreated utilizing the recommendations provided throughout this book. This appendix provides examples of the categories of templates that any project/program manager can reference to be more effective in their program/project management techniques. We have created a robust set of templates in our Library Manager application available at My Digital Libraries. The application allows any organization to customize templates proven to be effective by the author who has managed multiple sized projects in both the private and public sector over the last 30 years. There is also a subscription service that allows you to download templates as they change or new ones are developed over the years.

Category	Description
0000	**Program Infrastructure Setup** to include Collaboration Site structure and content and a library with standard templates.
1000	**Business Development** to include building new opportunities and structuring and managing the pursuit team.
2000	**Recruitment Management** to include sourcing high impact candidates and managing the recruiting effort.
3000	**Integration Management** to include delivery communication and continuous improvement for efficient solution delivery.
4000	**Communication Management** to include both external communication and internal team communication.
5000	**Requirement Management** to include new project work intake and defining definition of 'done'.
6000	**Solution Development** to include standard documents, migration, risk, and quality solution management.
7000	**Security Management** to include monitoring and remediation standard operating procedures.
8000	**Information Dissemination** to include visualization, dashboards, reports, and data extract and migration solutions.
9000	**Operation and Maintenance** procedures including production support and customer support components.

Appendix 4: Index

Appendix 5: About the Author

Stephen T. Boschulte is a successful information management specialist helping Fortune 500 companies maximize the return on the information they collect. Mr. Boschulte obtained his Bachelor of Science in Computer Engineering from the University of Notre Dame and his Master's degree in Business Administration from the University of Virginia. Upon completion of his studies, Mr. Boschulte spent over 28 years working at large consulting companies including Ernst & Young, Cap Gemini, Deloitte Consulting LLP and CACI. Over those years he obtained his Project Management Professional (PMP) certification (2001), Certified Information Systems Security Professional (CISSP) (2011), and Project Management Institute's Agile Certified Practitioner (PMI-ACP) (2013). Mr. Boschulte has managed teams on both small and large technical solutions to over twenty Fortune 500 companies and large government organizations. His primary area of expertise is Information Management. In 2010, Mr. Boschulte published his first book, A Practical Guide for Implementing an Enterprise Information Management Program. Since then, he has focused on helping organizations become more efficient by implementing scaled Agile life cycle development and structuring the program infrastructure required to optimize successful high-performance teams that best position the organization for continuous improvement using a standards-based approach. This book is a cumulation of leading practices garnered and created from his experience in these organizations.

Mr. Boschulte welcomes your comments and can be reached at sboschulte@OHPTeams.com

Bibliography

(2024, December). From National Institute of Standards and Technology: https://www.nist.gov

About. (n.d.). From International Information System Security Consortium: https://www.isc2.org/about

About. (2024). From Scaled Agile Framework: https://scaledagileframework.com/about/

Agile Glossary. (2024, December). From Agile Alliance: https://www.agilealliance.org/glossary/backlog/

AHA. (2024). *AHA Certification Center*. From American Hospital Association (AHA): https://www.aha.org/career-resources/certification-center

ARMA. (2024). *Frequently Asked Questions*. From Association of Records Managers and Administrators: https://www.arma.org/page/FAQ

Atlassian. (2024). *Agile/Kanban*. From Atlassian: https://www.atlassian.com/agile/kanban

Atlassian. (2024). *Kanban vs Scrum*. From Atlassian: https://www.atlassian.com/agile/kanban/kanban-vs-scrum

Axelos.com. (2021, September). *An Overview of the ITIL Maturity Model*. From Contentstack.com: https://eu-assets.contentstack.com/v3/assets/blt637b065823946b12/bltc4d875b75a1442ce/618029daa0038563ae7fdbd6/An_Overview_of_the_ITIL_Maturity_Model.pdf

BECNT. (2023, 02). *Sample Internal Audit Report for QMS*. From Business Excellence Consultancy: https://becnt.com/wp-content/uploads/2023/02/Sample-Internal-Audit-Report-for-QMS.pdf

Boschulte, S. (2010). *A Practical Guide for Implementing an Enterprise Information Management Program.* Minneapolis: London Street Press.

Business Winning Tips: Color Team Reviews. (n.d.). From Shipley Associates: https://www.shipleywins.com/blogs/color-team-reviews

Cooks, A. B. (2022, December 13). *Don't Salaries Matter?* From Standof Social Innovation Review: https://ssir.org/articles/entry/dont_salaries_matter

D.H. Maister, C. H. (2001). *The Trusted Advisor.* Simon & Schuster.

Digital AI. (2024). *17th State of Agile Report.* From Digital AI: https://digital.ai/resource-center/analyst-reports/state-of-agile-report/

Glossary. (2024, December). From Computer Security Resource Center: https://csrc.nist.gov/glossary/term/zta

Goleman, D. (2000, 03). *Emotional Intelligence: Leadership That Gets Results.* From Harvard Businesss Review: https://hbr.org/2000/03/leadership-that-gets-results

Home . (2024, December). From Shipley Associates: https://www.shipleywins.com/

ISACA. (2024). *Appraisals.* From CMMI Institute: https://cmmiinstitute.com/learning/appraisals

ISO. (2024). *ISO 9000 Family: Quality Management.* From Internaltional Standards Organization: https://www.iso.org/iso-9001-quality-management.html

ISO 9001:2015. (2015). *Quality management systems - requirements.* From International Standards Organization: https://www.iso.org/standard/62085.html

ISO/TC 176. (2024). *Quality Manaagement and Quality Assurance.* From International Standards Organization: https://committee.iso.org/home/tc176

ITIL. (2024, 12). *What is ITIL Service Management.* From Axelos: https://www.axelos.com/certifications/itil-service-management/what-is-it-service-management

Jobs Openings and Labor Turnover Statistics. (2024). From Bureau of Labor Statistics: https://www.bls.gov/news.release/pdf/jolts.pdf

Knight, R. (2024, 4 9). *6 Common Leadership Styles - and How to Decide Which to Use When.* From Harvard Business Review: https://hbr.org/2024/04/6-common-leadership-styles-and-how-to-decide-which-to-use-when

Landry, L. (2020, January 14). *How to Delegate Effectively: 9 Tips for Managers.* From Harvard Business School Online - Business Insights: https://online.hbs.edu/blog/post/how-to-delegate-effectively

Lean Six Sigma. (2024). *Advanced Product Quality Planning (APQP).* From Lean Six Sigma Definition: https://www.leansixsigmadefinition.com/glossary/apqp/

Lietz, S. (2016, June 05). *Blog <- Shifting Security to the Left.* From DEVSECOPS: https://www.devsecops.org/blog?author=54dc6220e4b085335dd9d630

Liuba. (n.d.). *Seven Stages of Grief.* From Psychology Tips: https://psychology.tips/7-stages-of-grief/

263

Mayo Clinic Staff. (2023, November 21). *Healthy Lifestyle*. From Mayo Clinic: https://www.mayoclinic.org/healthy-lifestyle/stress-management/in-depth/positive-thinking/art-20043950

NIST Computer Security Resource Center. (2024, December). From NIST: https://csrc.nist.gov

NIST Information Technology Laboratory - Computer Security Division. (n.d.). From National Institute of Standards and Technology: https://www.nist.gov/itl/csd/computer-security-resource-center

NIST IR 8397 Guidelines on Minimum Standards of Developer Verification of Software. (2024). From National Institute of Standards and Technology: https://csrc.nist.gov/pubs/ir/8397/final

PMI . (2024). *Pulse of the Profession 2024*. From Project Management Institute: https://www.pmi.org/learning/thought-leadership/pulse/future-of-project-work

PMI. (2012). *The Project Management Office - In Sync with Strategy*. From Project Management Institute: https://www.pmi.org/-/media/pmi/documents/public/pdf/white-papers/value-of-pmo.pdf?v=9a263118-2d42-4d5c-ae60-434ee2d7854e

PMI. (2024). *Certifications*. From Project Management Institute: https://www.pmi.org/certifications

Porter, M. (1979, March-April). Harvard Business Review. *How Competitive Forces Shape Strategy*.

Rebecca Joy Stanborough, M. (2019, October 15). *Benefits of Reading Books: How it Can Positively Affect Your Life*. From Healthline: https://www.healthline.com/health/benefits-of-reading-books

S, S. (2010). *Project and process integration: how to usefully combine two work management models*. From Paper presented at PMI®

Global Congress 2010—EMEA, Milan, Italy. Newtown Square, PA: Project Management Institute.: https://www.pmi.org/learning/library/project-process-integration-combine-models-6801

www.ingramcontent.com/pod-product-compliance
Lightning Source LLC
Chambersburg PA
CBHW070305200326
41518CB00010B/1904